Landscapes of Settlement

In spite of the contemporary dominance of urban-based lifestyles, over half the world's population remains rural, a proportion that rises to two-thirds in the less developed regions of the globe. Even today's most highly urbanised societies grew from rural roots – settlements that are centuries old forming the framework for modern life.

Landscapes of Settlement discusses the role and importance of rural settlements, emphasising their historical impact on present-day society as well as their contemporary role in the developed and developing world. Analysing the language of settlement and questions of definition, the book looks at a series of contexts within which settlements can be explored and explained. Beginning with the basic ingredient of all settlement – the human dwelling – the author traces the aggregation into farmsteads, hamlets and villages and the ways these associate, together with towns, to form definite patterns of settlement.

With copious global case studies, maps and models, *Landscapes of Settlement* presents and defines the principal processes at work within rural settlement systems, emphasising the role of time as the matrix within which these processes operate at varying levels of intensity throughout the world.

Brian K. Roberts is Reader in Geography at the University of Durham.

Charles Seale-Hayne Library
University of Plymouth
(01752) 588 588
LibraryandITenquiries@plymouth.ac.uk

Paddy O'Sullivan
Reading Lists

If you choose to believe me, good. Now I will tell how Octavia, the spiderweb city, is made. There is a precipice between two steep mountains: the city is over the void, bound to the two crests with ropes and chains and catwalks. You walk on the little wooden ties, careful not to set your foot in the open spaces, or you cling to the hempen strands. Below there is nothing for hundreds and hundreds of feet: a few clouds glide past; farther down you can glimpse the chasm's bed.

This is the foundation of the city: a net which serves as passage and as support. All the rest, instead of rising up, is hung below: rope-ladders, hammocks, houses made like sacks, clothes-hangers, terraces like gondolas, skins of water, gas jets, spits, baskets on strings, dumb-waiters, showers, trapezes and rings for children's games, cable-cars, chandeliers, pots with trailing plants.

Suspended over the abyss, the life of Octavia's inhabitants is less uncertain than in other cities. They know the net will last only so long.

Landscapes of Settlement

Prehistory to the present

Brian K. Roberts

London and New York

And the days are not full enough
And the nights are not full enough
And life slips by like a field mouse
Not shaking the grass
(Ezra Pound)

First published 1996
by Routledge
11 New Fetter Lane, London EC4P 4EE

Simultaneously published in the USA and Canada
by Routledge
29 West 35th Street, New York, NY 10001

© 1996 Brian K. Roberts

Typeset in Palladia by Solidus (Bristol) Limited
Printed and bound in Great Britain by
Clays Ltd, St Ives PLC

British Library Cataloguing in Publication Data
A catalogue record for this book is available from the British Library

Library of Congress Cataloguing in Publication Data
A catalogue record for this book has been requested

✓ ISBN 0–415–11967–7
0–415–11968–5 (pbk)

Contents

Plates

The following plates appear between pp. 86 and 87.

1   A traditional Bedu goatskin house
2   The modern settlement of Safawi, north-east Jordan
3   The village of Keyala, Equatorial Province, Sudan
4   The village of Wiatua, Equatorial Province, Sudan
5   Sura Village, Jos Plateau, Nigeria
6   Village pond at Besser, Samsø, Denmark
7   Staindrop, County Durham, England
8   Excavated interior of the church at Wharram Percy, East Riding, Yorkshire, England
9   Roadside view of Thoralby, North Riding, Yorkshire, England

Figures

Introduction

This book has been written to try to convey to others something of the curiosity and excitement I feel about rural settlement. Present settlement landscapes are not in any way static. They are moving through time. They come from a past about which we understand something, through a present that is bewildering in its diversity and complexity, and will pass to a future which is, thankfully, hidden. Thus, to understand rural settlements, historical explanations are needed, involving the reconstruction of past events by means of the unique phenomena resulting from these events. However, such conclusions, perhaps evaluations as much as conclusions, are not reducible to the consequences of natural laws. Explanation must be sought in the realms of contingent detail, the logical association of observations and linking arguments and models.

Settlement landscapes are signs and symbols of centuries of human endeavour. Much of this has comprised a blind, unremitting, often grinding toil which has been, and indeed still is, the lot of most of humankind: nevertheless, earlier societies were able to generate works of the mind which still have the power to stir – the *Cattle Raid of Cooley*, the epic of *Beowulf*, the poetry of Dafydd ap Gwilym, Shakespeare to choose only a few cases drawn from these islands – as well as vast changes in cultural landscapes. This was all done, until the last century, amid conditions where doctoring was non-existent or rudimentary, life expectancy limited, pension schemes absent, and warfare and violence endemic. In the ordering of settlement and economic life, above all in the concept of sharing resources, however unevenly, imagination and ingenuity can be seen to be at work, coupled with a deep understanding of the physical environment.

Two further points must be made: first, the maps and diagrams included in this book are not merely adjuncts to the text, they are an integral part of the argument; they must be studied in relation to the text. They are designed to extend across varied scales of investigation, and while they nominally illustrate particular chapters, many cross-references will be found *between* chapters, while for the varied scales examples have been selected to cross-reference with each other; in short, there is much more that could be discussed concealed within the drawings. Second, in all writing there is a need to balance generalisation and detail: this is a key problem when writing on rural settlement. Each chapter will be found to contain some 'general principles', points to be carried, produced and used in a variety of circumstances: the supporting examples carry more detail, often by no means fully explored, concerning real cases. A word of warning here: reality is vastly complex, and compressed accounts are notoriously distorting lenses through which to see reality.

There are inevitable omissions; undoubtedly a major one is to be found in the essential systems which support and sustain the rural settlements discussed, most clearly seen in the surrounding field systems, but extending to those less visible aspects, husbandry practices and farm structure, landownership and marketing arrangements; no one study can deal with all aspects, and the intention here is to show ways in which the visible, observable aspects of settlement, from farmstead to village, can be approached. This, of course, leads to another omission, the village to town transition.

Nevertheless, this study retains an essential core: in order to understand the rural landscapes of the present, anywhere in the world, it is necessary to seek understanding via the past, for as is stressed repeatedly, the present is rooted in the past. These words are being written in a farmhouse, probably of late-eighteenth-century date, but the Norman keep of Brough castle lies a few dozen metres away, the great stone keep constructed after its precursor was burned during an attack by William the Lion, King of Scotland, in the year 1174, when six knights defended the wooden tower on the motte until they were in danger of being burned alive. Below, half a mile or so away, lies Market Brough, a small town plantation, now no more than a bypassed village, but probably established when the English recovered the area and built the stone keep, all before 1200. However, the drumlin on which the present castle sits had attracted the eyes of the military engineers of the Roman army, who established the fort of *Verteris* on the supply line between the great base at Catterick and the frontier town of *Luguvalium*, Carlisle; the soldiers' bath house lies beneath the farmyard, but the task which brings me to the area involves tracing the successive layers of occupation between the prehistoric period and the establishment of the medieval villages which give structure to the present landscape. The bright low sunlight of a January anticyclone has shown me wonderful things, landscapes superimposed upon landscapes. This is the excitement of settlement.

CHAPTER 1

Settlement landscapes

This is a book concerned with rural settlement, i.e. that part of the settlement hierarchy most closely concerned with working the land. While no study of limited dimensions can claim to be worldwide, it adopts a viewpoint which draws examples from regions as widely separated as the tropics and the temperate zones and from several continents. Further, it recognises that past conditions are an essential part of explanation, so that any study involving processes of change must view rural settlements within a time framework. Finally, it builds around a series of diagrams and maps. These are used both as a means of handling the varied scales, ranging from worldwide to the internal structure of a single house, and as a means of generalising through distribution maps and settlement and varied models. By these means both time and space can be compressed into a single page.

Rural settlement is always experienced through the particular. Focus for the chapters which follow can be drawn from three descriptions of actual settlements, beginning with John Hunter's description of a single house in northern Ghana, examined in 1967 (Figure 1.1):

> The basic settlement unit is the house. In outward appearance it resembles a miniature fortress, but internally it may be likened to a honeycomb. Certain architectural conventions are rigidly observed. The entrance always faces west. 'Because it is forbidden to build otherwise' was the usual response to Hunter's queries on this point. There appears to be no traditional explanation for this orientation, although ... all entrances were situated in the lee of the line squalls which advance from east to west across the area in each rainy season. Entrance to a house is always through the cattle-yard where the farmer's wealth, his manure, accumulates. Next there is the main granary, an imposing cone-like structure some ten or twelve feet tall. Facing the granary, the cattle-yard, and the gateway is the room of the head of the house. This arrangement never varies. Around these key features, the house develops, not haphazardly, but in accordance with strictly observed conventions. Each wife is entitled to her own room, kitchen and courtyard, so that the physical layout of the house directly reflects family

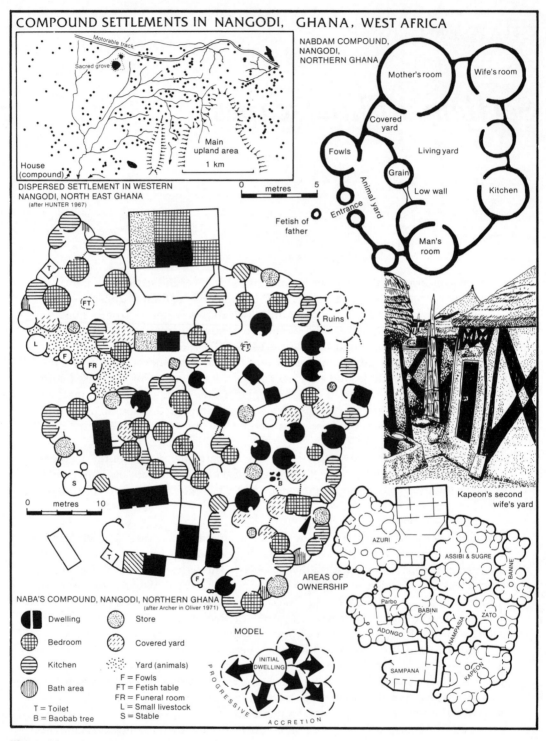

COMPOUND SETTLEMENTS IN NANGODI, GHANA, WEST AFRICA

Motorable track

Sacred grove

House (compound)

Main upland area

1 km

NABDAM COMPOUND, NANGODI, NORTHERN GHANA

Mother's room

Wife's room

Covered yard

Fowls

Living yard

Grain

Low wall

Kitchen

Animal yard

Entrance

Fetish of father

Man's room

DISPERSED SETTLEMENT IN WESTERN NANGODI, NORTH EAST GHANA
(after HUNTER 1967)

Ruins

T
FT
L
F
FR
FT

S

B

0   metres   10

T

F

Kapeon's second wife's yard

AREAS OF OWNERSHIP

AZURI

ASSIBI & SUGRE

BANNE

Parbo

BABINI

ZATO

ADONGO

NAMPASIA

SAMPANA

KAPEON

0   metres   5

NABA'S COMPOUND, NANGODI, NORTHERN GHANA
(after Archer in Oliver 1971)

- Dwelling
- Bedroom
- Kitchen
- Bath area
- Store
- Covered yard
- Yard (animals)

F = Fowls
FT = Fetish table
FR = Funeral room
L = Small livestock
S = Stable

T = Toilet
B = Baobab tree

MODEL

PROGRESSIVE

INITIAL DWELLING

ACCRETION

Figure 1.1

development. Society is patrilocal, thus when a son marries, his wife too requires a courtyard or domestic sub-unit in the family house. As a family expands, its house expands, domestic sub-units being added, biologically rather like cellular reproduction. Senior wives' quarters flank the room of the head of the house; junior wives, married sons, and other members of the family occupy rooms added to the rear. Large houses are a veritable maze of courtyards.

(Hunter in Salter 1971: 171–2)

One dwelling Hunter surveyed contained no less than seven courtyards, yet he described it as 'not large'. The term *compound* is applied to such an aggregated dwelling, and in Figure 1.1, while the smaller of the examples is derived from Hunter's work, the larger, Naba's compound, drawn from another source, indicates the way in which twelve complex households are drawn together within one cellular homestead, whose growth is summarised in a simple model. Of course each of these compounds sits amidst its home-farm, whose fertility is maintained by applications of the manure, spread radially outwards from the dwelling, so that a sharp fertility gradient develops with the better-quality cropland lying near the house. 'Space permitting a home-farm is circular in shape. Fully circular farms are thus to be found where population density is low, as among new settlements, on the periphery of older, more densely populated areas' (ibid.: 169). When a man dies, the home-farm is divided, often with a younger brother being forced to establish himself on new land, but always locating his new compound as near as possible to his father's grave and the domestic and lineage shrines. The house or compound is more than a shelter or store, it is an expression of culture and economy.

A second wholly contrasting description is drawn from two accounts, both originating in the 1950s, concerning mining villages in the north of England:

The new villages [Figure 1.2] ... which were built around the pit heads, often on the hill tops where the upper seams outcropped or came near the surface, were dreary, soulless places. They consisted of long, parallel, monotonous rows often with unmade streets, primitive sanitation and there was almost a complete lack of social facilities. They were built as quickly and cheaply as possible and as close to the mine as possible. Little or no thought was given to the real needs of the inhabitants.

(Durham County Council 1951: 18)

They were not so much towns as barracks; not the refuge of a civilisation but the barracks of an industry. This character was stamped on their forms and life and government ... they were settlements of great masses of people collected in a particular place because their fingers or their muscles were needed on the brink of a stream here or the mouth of a furnace there. These people were not the citizens of this or that town, but the hands of this or that master.

Sid Chaplin, a miner by birth and author of the second description, hammers this point home as if he were drilling a shot hole:

The villages were built overnight – the Americans are much more realistic about mining than we are. They know it's a short-lived thing relatively speaking ... so they talk about mining camps; we talk about villages, which is one of the oldest words in the language. It means a permanent settlement. But most of the Durham villages were camps. The first street was Sinkers Row, and an upper and a downer.

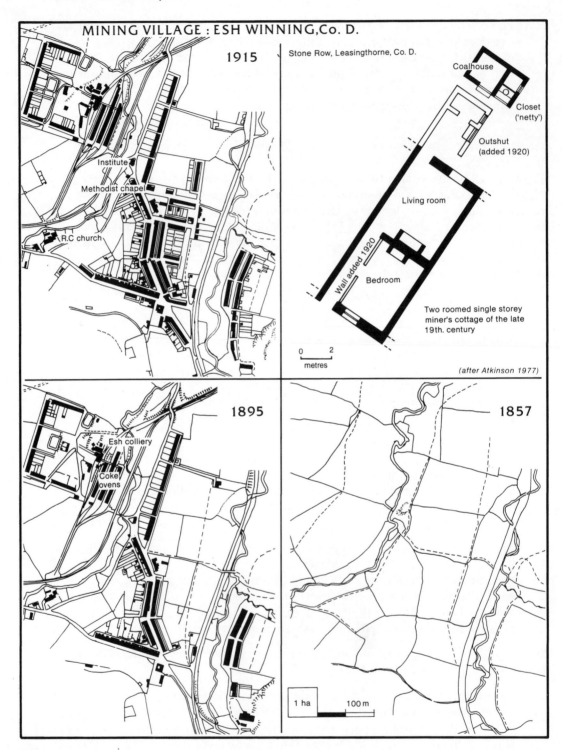

Figure 1.2

At best a sort of but-and-ben arrangement. That is a little scullery with a fairly large kitchen and if you wanted a bedroom for the kids, you put planks across the rafters.

(in Bulmer 1978: 59–82)

These images of Durham mining villages emphasise three things: first, the interlocking of settlement and life; indeed Sid Chaplin's contribution continues with a lengthy account of the social role of the 'netty' – an outside earth closet or toilet, placed in the backyard – concluding that in addition to reading and studying, 'a lot of praying and sermon composing went on in the netty. They played a very large, a very vital part, quite apart from the hygiene function, in the lives of the communities.' He is joking and exaggerating a little, but what he says is more understandable when it is grasped that within the house itself large families, sometimes with as many as seven, eight or even more, shared very limited room space: 'two rooms in an upper and downer and one room with loft space and a smaller one attached'. Second, the account touches not only the character of the mining village as it now is, but seeks understanding within time, for each village was first planted and then grew. Often beginning as a single row of terraced houses – Sinkers Row, associated with sinking the shaft – it was then expanded, generally a few houses at a time, a row here and a row there. Northern English mining villages were often, quite literally, a field of a dozen or so acres given over to closely packed houses rather than grass. Over the hedge their claustrophobic pseudo-urbanism normally gave way directly to fields still under agricultural use. Third, Chaplin's use of the term 'barracks' is a chilling reminder of the essentially ephemeral nature of these settlements: once the mine closes the economic *raison d'être* of the settlement is undermined, and the least viable villages, those places where planning rules dictated that investment should be limited to the life of the existing property, often became wholly depopulated, a generation of deserted villages; field to field in just over one hundred years. The village of Esh Winning (Figure 1.2), its very name speaking of its mining origin, has in fact survived to the present because of its size and good communication links, but the sample house from Leasingthorne is from a settlement that is now substantively depopulated.

The contrast between the two settlements could hardly be greater. Indeed, were it not for its small size and isolation within the countryside, the mining village would challenge inclusion within a study of rural settlement. Nevertheless, the contrast is instructive: the African settlement, truly rural, is bonded directly to the land and supported by its produce, while the mining village, drawing its sustenance from the coal seams deep below, grows upon it. The mining village is the living space of a group of people drawn together by the need to work, heterogeneous in character, integrating into a single community people of diverse origins who – at least in the first stages of growth – were immigrants drawn from other equally fragmented settlements. The African settlement consists of people linked by polygamous marriage and kinship. One is wholly rural while the other is industrial and apparently has no place within a study of rural settlement, but it does serve as a reminder that rural settlements may derive livelihoods other than from arable fields. Indeed in western Europe many contemporary villages of wholly rural origin are often now completely divorced from their rural base and origins. To create too narrow a definition of the term 'rural' would be to ignore much that is patently part of the countryside.

A third and final settlement description will not at first be placed within a context of time or space:

> All but the richest families lived in low houses thatched with turf, heather or reed, built wholly of timber or perhaps more commonly of turves and rubble fitted around a timber frame. At night and in winter the most valuable beasts occupied one end of the dwelling, the humans of both sexes and all ages finding what living space they could at the other end.

The character of the environment around the settlement is sketched in by the author as follows:

> There would have been an immediate and continuous assault upon the nose in what was quite literally a stinking country: the warm homely smells of cattle, horses, and hay, of food cooking in stone ovens or on open hearths, mingling unavoidably with the acrid reek of peat fires, the putrid odour of rotting meat and fish or of untreated skins and hides, and in every inhabited locality the stench of animal and human ordure.
>
> (Barrow 1981: 20–1)

This description would be true of many parts of the world today, but its author, the historian Geoffrey Barrow, begins his account of the years between 1000 and 1300 with a magical description:

> Scotland in our period was, for a start, a wonderfully quiet country, where nearly all the sounds would have stemmed from natural causes like weather and running water, from farm animals, the barking of dogs, the shouts of village children, or the songs of men and women as they worked. The loudest artificial sound familiar to ordinary people may well have come from church bells. Moonlight and starlight would have been much more keenly appreciated than they can possibly be by a generation that takes electricity and the sadly inescapable sodium lamp for granted.

This description, with minor modifications, would have been appropriate for the peasant settlements found throughout northern Europe at that time. Nevertheless, such simple peasant communities should not be underestimated, for these were part of societies which generated sufficient skill and wealth to build the great stone castles, cathedral churches and abbeys so essential to the modern tourist trade – in Scotland at Kirkwall, Arbroath and Dunfermline, Jedburgh and Melrose, but also, of course, the Romanesque splendours further afield, the Tower of London, the cathedrals at Durham, Worms, Bayeux, or the abbey of the Holy Trinity in Caen.

There is little in this description of medieval Scottish settlement that is, superficially at least, different from what can be seen in the plan of the eighteenth-century 'fermtoun' preserved as an open-air museum at Auchindrain (Fenton and Walker 1981: 20, 77). The physical character of the buildings (Figure 1.3) is glimpsed in the small drawing derived from a photograph taken of structures at Kinlochewe which may date from the late eighteenth or early nineteenth century. Such small clusters of dwellings, with human beings and cattle sharing the same roof – for the dwellings at Auchindrain and in the medieval phase of the hamlet at Jarlshof both allow for this – provided shelter and sustenance for a few families. They may have originated in the gradual accretion of new

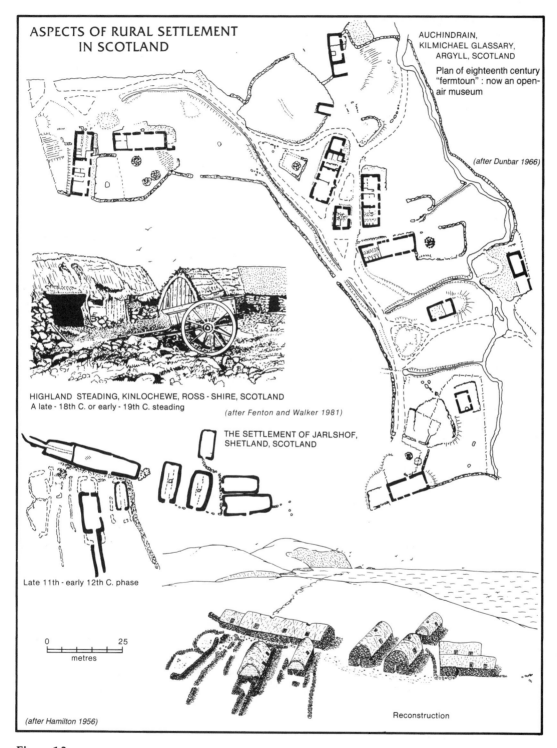

ASPECTS OF RURAL SETTLEMENT
IN SCOTLAND

AUCHINDRAIN,
KILMICHAEL GLASSARY,
ARGYLL, SCOTLAND

Plan of eighteenth century
"fermtoun" : now an open-
air museum

(after Dunbar 1966)

HIGHLAND STEADING, KINLOCHEWE, ROSS-SHIRE, SCOTLAND
A late-18th C. or early-19th C. steading

(after Fenton and Walker 1981)

THE SETTLEMENT OF JARLSHOF,
SHETLAND, SCOTLAND

Late 11th-early 12th C. phase

0   25
metres

(after Hamilton 1956)

Reconstruction

Figure 1.3

farmsteads, those of sons, other relatives or even foster children, to an initial parent unit. This was the probable origin of the small hamlet of Viking age seen at Jarlshof (Hamilton 1956), the largest building being the mother farmstead, to which daughter dwellings accreted, the homes of the original settlers' sons and their wives.

The north and west of Scotland, both mainland and islands, contain many regions in which arable land has always been in very short supply, indeed regions such that the arable lands could only be maintained in good heart by frequent back-breaking applications of manure, dung, seaweed and/or shell sand, carried in creels, or baskets, on the backs of the inhabitants. These informal hamlets, originally kin-based, tucked into small, favoured patches within a generally rather hostile farming environment, are characteristic of the northern fringes of Europe to this day. In these survivals we may get one glimpse of the possible origin of the village in regions with richer agricultural potential, a point to be examined in a later chapter.

These descriptions beg many questions. Not least, how is it possible to generalise about settlement, given the diverse factors which must be invoked to explain what is observed? Physical, social and economic factors all play a role, while the flat statement 'Because it is forbidden to build otherwise' raises deeper questions of psychology and beliefs. The three varied accounts speak of varied contexts for birth, marriage and procreation, shelter and storage, labour and death; each presents a challenge to think about a particular lifestyle and the way in which habitat and habitations interrelate. In all cases, the description moves, however briefly, inside the house, for house and yard form a nexus within which each culture expresses itself, is perpetuated, reiterated and changed through the lives of individuals and families.

Rural settlements: a general view

The themes which make up the study of settlement demand awareness that many scales of enquiry must be used: the lower section of Figure 1.4 uses a logarithmic scale against which to show some of these. The diagram extends from the interior of individual buildings, through associations of buildings and other features – plan elements – to the characteristics of individual settlement shapes – their morphology – and thence to the qualities of regional, and by extension – for the scale stops short at 10,000 square kilometres – national and eventually world patterns. Nevertheless, the physical characteristics and physical contexts of settlements only provide one possible component of study, which must also embrace cultural contexts, involving subtle and complex interrelationships, both within each individual settlement and between those similar or dissimilar settlements making up settlement patterns. By setting these within a framework which embraces increasingly large areas of terrain the lower portion of Figure 1.4 emphasises the extent to which scale contrasts, and the tensions between generalisations made at varied scales, must permeate all discussion. What is true at one scale is not necessarily true at another. Ultimately the logical limit of humankind's habitable environment is to be found in the physical size of the earth, so that studies of rural settlement must range from the whole earth to the internal details of individual dwellings.

Further, each of the three descriptions and associated maps and diagrams are permeated by an awareness of time. Each, indeed every, settlement possesses a temporal

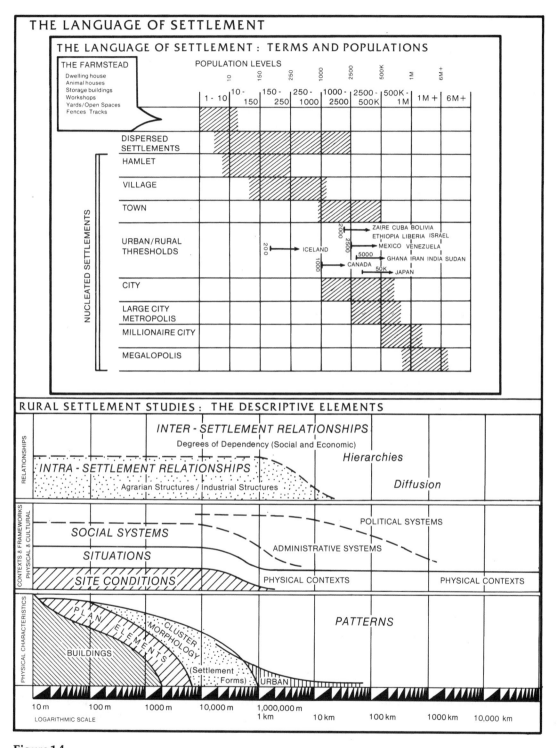

Figure 1.4

path or trajectory through time, a line of passage through years, decades and centuries, from past to present and on to future. This is happening to all the settlements, within a single region or a single country, and leads to a further important idea, that individual settlements possess a duration in time. There is a time when a given place did not exist, a time when it comes into being and then, permanent as the rural settlements of Europe or North America may seem within a brief human lifespan, a time when it has served its purpose and must inevitably pass out of existence as surely as did the encampments of simpler lifestyles. Should the physical structure of a settlement be of non-durable materials, of wood or thatch, and should the economy place such demands upon the land that rapid local exhaustion of soil or of grazing occurs, then habitations may never become permanent, even within the span of a single human lifetime. When Sid Chaplin described mining villages as 'camps' he was sharply aware that they would not last for ever and when a mine closes, destroying the settlement's *raison d'être*, a mining village must die unless another means of supporting the people is introduced. Thus it is that depopulated and deserted rural settlements are an important ingredient of most rural landscapes. In fact all settlement is a product of time and for this reason it is necessary to incorporate a measure of historical explanation into any study of settlement.

The attributes of rural settlements

Figure 1.5 is a model ultimately based upon work done in New Guinea by an anthropologist (Fraser in Brookfield 1973: 75–96), but in this version it incorporates the simplest type of rural focus, a steading from the Faroe Islands (Uldall *et al.* 1972: Building 15). Nevertheless, it is designed to be applicable in many contexts. The circularity invokes the seamless qualities of many of the factors which must be assessed when studying settlement; each runs into the other, each repurcusses upon the other. The circularity expresses this better than either a grid or a matrix, which both appear finite and bounded. The inner plate contains a basic listing of the key attributes, beginning, top right, with major goals, and progressing through economic base, locational regulators, size regulators (i.e. physical dimensions of each settlement and its productive land), population regulators, through decision-making contexts and enforcement, to the summation of many forces, either centripetal or centrifugal, which eventually lead to nucleation or dispersion. A second circle, amplifies the key points, while a third circle provides a simple categorisation of all the factors involved, economic, social and political. It will be noted that physical factors are present but never isolated, for they are always evaluated through a cultural filter. In this model they appear under locational and size regulators, but could equally be extended, for example, to population regulators. Marshland settlements were, until they were tamed and drained, notoriously subject to fevers such as malaria (Braudel 1972: 62–75), conditions as much a product of the physical environment as are the factors of site and situation, to be seen in Figure 2.6. Such factors cannot be ignored, but within a developed settlement system they are rarely wholly central, and Figure 1.5 places them in a more realistic, broader perspective. Of course, in difficult environments, such as regions where water is in limited supply, then the influence of water rights and access to water must be correspondingly enhanced.

There is little need to discuss this self-explanatory diagram in detail, but two particular points should be noted: first, the ultimate goal of all rural settlements through

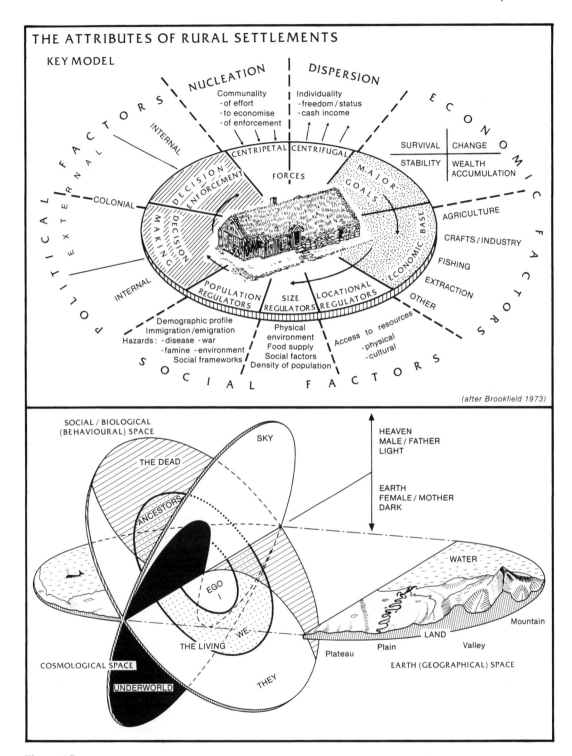

Figure 1.5

all time and in all places has been survival. As much can be said of any human association, be this a family, a settlement, a firm, a country or a culture. Survival may, however, involve seeking either stability or commitment to willed change, i.e. change that is brought about by internal factors and is not the result of externally imposed accidents of nature or war. It is sobering to reflect that most human beings have lived their lives in contexts in which sustaining the status quo was a dominant concern; indeed this has been a key characteristic of *rural* society, dependent upon the annual cycle of the agricultural year or, before that, the passage of seasonal events in the world of vegetation and animals. A fourth major goal is wealth accumulation. This is an objective which arises in developed capitalist societies once the immediate problems of survival have been solved, when a belief in logically willed change becomes a guiding principle. The social disintegration linked with this is an observable fact of life (Gellner 1988: 53–67), but in fact, survival is more tenuous than most of us appreciate, and within the last decade, for reasons right or wrong, an awareness of 'environmental issues' is raising uncomfortable questions. Second, decision making and decision enforcement (Figure 1.5), both *internal*, arising within each society – the impositions from governments and local rulers, priests and planners – and *external*, introduced or imposed from outside, affect the character and qualities of rural settlements. In the recent past external forces were often represented by 'colonial' powers, but increasingly controls are introduced through agreements voluntarily entered by nation states, as in the European Community. In this summary diagram the distinctive qualities of nucleation and dispersion, the concentration or scattering of settlement entities, are introduced as two sectors amid the whole, and seen as the resultant of all the attributes operating through time.

Perceived space

The lower section of Figure 1.5 moves the argument into less concrete realms. It suggests that all individual persons – 'ego', at the centre – possess complex relationships with the space which surrounds them and forms a framework for their life. The first, geographical or earth space, consists of the elements of the natural environment – the physical elements of land and sea, plain and mountain. The causal links between earth space and rural lifestyles have been the subject of much discussion and analysis by scholars. However, human beings have envisaged worlds which have other qualities, other dimensions, the worlds of the mind. Ultimately all of these are concerned with the trajectory of the individual from birth, through childhood to maturity and awareness, to death and the question-mark beyond. Some aspects of the myths generated by the mind are built into the two further dimensions of space shown in this model, behavioural space and cosmological space. The former is concerned with social relationships between individuals, and between the individual and the group and the group and the individual – this choice of words being more than mere pedantry – while the latter concerns broader beliefs and attitudes, normally focused within each society's myths and religious beliefs. In a nutshell, thought is as powerful as any physical factor.

The diagram, admittedly with an element of tongue in cheek, is built around the ancient Chinese symbol of *yin* and *yang*, one particular aspect of those symbolic oppositions which permeate thought and myth, concerned not least with male and female forces. Twentieth-century rational western society takes little account of such factors for they cannot be included in cost–benefit analyses, but many parts of the world, even economically successful portions, still utilise such intangible contributions. For instance, a Chinese businessman, in Hong Kong, will still use a Feng-Shui compass for determining the most propitious location for an office building (Oliver 1987: 167–70; Skinner 1982). This comprises a set of movable plates on which are inscribed astrological signs and is an instrument developed as early as the tenth century AD bringing together different forms of geomancy. Introduced into Japan during the seventeenth century, Feng-Shui was interwoven with the practicalities of what in England would have been called 'improvement', i.e. the rationalisation of agriculture and improving both farm structure and the irrigation of rice fields, a process which also involved both the lateral movement of old villages to new sites – sometimes by consolidating two together – and also creating wholly new planned settlements with checkerboard plans (Singh 1972: 251).

This is a difficult topic to approach and sketch briefly. The western mind is culturally conditioned to be sceptical, yet in the broader context of human experience the cold concept of 'economic man' must also be questioned. On the other hand, it is too easy to descend into mystical nonsense and it is clear from the generalised example from Japan that more than geomancy could be involved. One thing is quite clear: human beings need more than physical and economic space; individuals and groups also require psychological space, which involves values, beliefs, symbols and meanings – here one only has to think of the vast amounts of energy and blood expended on achieving 'nationhood'. In an Indian context the deeply embedded caste system is alien to the western mind, yet, as Spate noted, has met the needs of a large fraction of humankind for many centuries. There are, or at least were, four great caste groupings: Brahmins (priests), Kshatriyas (warriors or rulers), Vaisyas (traders) and Sudras (cultivators), together with lower social groups known variously as Untouchables, Scheduled or Exterior Castes. Divisions extend to such specific tasks as those undertaken by washermen (Dhobis), tanners (Chamars) and scavengers (Doms), and even distinctions based upon those who make black or white pots. Legislation has now mitigated some of the worst aspects of these attitudes, but the system originated in a system which once provided a place for everybody, for the intrusive conqueror and the aboriginal conquered (Spate 1957: 136–40). Values and beliefs, often conveyed through symbols, are, and always have been, expressed in a multitude of ways, although all find a grounding in that individual journey through life which is the common human experience. They are deeply important when appraising rural settlement, indeed all settlement, because the built environments, settlement forms and even settlement patterns created by human societies extend into, and are extensions of, these psychological realms – echoes of the statement to Hunter cited earlier, 'Because it is forbidden to build otherwise'.

The framework of the study

Numerous specific examples are incorporated into this study, but the framework adopted is essentially thematic – as the simple chapter headings show. Within this framework scale changes are used, both as a way of illuminating the diverse approaches needed for the investigation of rural settlement and as a way of emphasising the role of historical factors in generating the settlement systems now present. Chapter 2 – 'Nucleation and dispersion' – outlines the language of settlement, examining the meaning, or more properly meanings, of some of the most basic terms used, farmstead, village, nucleated, dispersed, pattern and form. Chapter 3 – 'Explanatory contexts' – begins at the scale of the state, presents three national distributions, and then examines the broad contexts within which such complex patterns can be discussed and explained, concluding with a well-documented study of the Netherlands. A sharp alteration in scale follows, and the next three chapters – 'House and farmstead', 'Settlement forms' and 'Settlement patterns' – move from the smallest objects of study, through the rich diversity of settlement plans, to local and regional conditions visible in contrasting patterns. The study concludes with a view of settlement systems on a world scale.

CHAPTER 2

Nucleation and dispersion

Discussions of the physical character of settlement tend to be obscured by questions of terminology (Adams 1976: 47–91). There is no escaping the fact that, as in any other subject, a certain amount of jargon must be used to express observations and ideas cogently. Hunter's African 'house', for instance, could equally be termed a 'farmstead', for in addition to several buildings for human habitation, it contained outbuildings, storage and work areas. The term *compound*, meaning an 'enclosure', conveniently indicates that it is different from European or North American farmsteads and is in fact made up of many elements which in other societies could be differently arranged or separated. The single farmstead takes many forms and may appear either singly or in groupings of varied sizes, but, nevertheless, it is the fundamental unit of both subsistence and commercial agriculture. From the farmstead, however this be defined, the farmer, sometimes with the members of his immediate family and sometimes with hired hands, works the surrounding lands. There are instances, particularly in the tropics, where fields can be cleared, planted and then simply left to reach a harvestable condition, while some farmsteads in temperate lands have become mere machinery and product stores, visited by workers. It is nevertheless generally true that farmstead and fields are bound together by the need to cultivate the land or nurture beasts. This farmstead–land relationship takes many forms and a definition of some of the basic terms is necessary.

Basic terms: the language of settlement

The words 'town', 'village' and 'hamlet' are commonly used to describe settlements throughout the world but they are extraordinarily difficult to define with precision. Indeed it must be stressed that in practice it is almost impossible to impose any absolute consistency of usage upon any of them (Adams 1976; Uhlig and Lienau 1972; *OED*). Key elements in the vocabulary of settlement are included in the following account (Figure 1.4).

The general term *farmstead* characterises the grouping of agricultural buildings forming the focus of an individual farm enterprise. Normally there will be a dwelling house, buildings for animals, storage for fodder and crops, working areas, yards, pens and open work-spaces, all delimited by fences, walls and access ways. There may, of course, be more than one dwelling and/or accommodation for family members or hired hands. Should the dwelling house be absent, then the term farm *offices* appropriately describes what is left, using the word 'office' in an older sense of 'the parts of a house used for household work or service' rather than dwelling. In Scotland the total complex would be described as a *steading*, while for the African example already cited the term *compound* is appropriate. In Britain and North America the term *homestead* is often applied, useful because it emphasises its social value, the place where family life takes place, where children are reared and learn basic skills. Two points should be emphasised: first, farmsteads vary tremendously in character and here the adoption of specific foreign-language or local dialect terms is a useful way of expanding and defining shades of meaning. Thus the Spanish *hacienda* has a specialist use, describing the large farmstead and house which are the focus of a great estate in Latin America, while, in sharp contrast, *onstead* is an appropriate northern English dialect term for a small farmstead whose small area of improved meadow and pasture land is set amid the open moorland of the Pennines. In this way the vocabulary can be extended with both grace and precision. Second, such qualifications are important, because while 'farmstead' will serve on most occasions, the diversity of types found within the real world ultimately necessitates the recognition of many subtleties. Where elements of joint tenure are present, i.e. where farmers are not wholly independent from each other but share lands and burdens, such as payment of rent in whatever form this takes (work, crops or money), then a joint farmstead or group farmstead can be identified, i.e. there can be three or more farmsteads but only one farm or worked land area. In other circumstances, in parts of South East Asia for example, a single building or longhouse can accommodate more than one family or even a whole community.

Hamlet is a neutral term for settlement comprising a cluster of six to eight farmsteads. In Scotland the term *fermtoun* has been used, normally for a group of from two to eight farmsteads, but in Ireland this would be called a *clachan*, in Brittany a *ker*, in Spain an *aldea*, in northern Germany a *Drubble* and more generally in Germany a *Weiler*. Early Welsh medieval laws provide a vivid picture for they tell us that a 'lawful *tref*' (yet another word for a hamlet) comprised 'nine buildings and one plough, and one kiln and one churn, and one cat and one cock and one bull and one herdsman' (Homans 1941: 26–8). At this period the plough was presumably pulled by a single large shared team of oxen and tilled only a limited amount of arable land, with other animals, under the control of the herdsman, using the communal grazings. The kiln was vital for baking bread, while the cock and bull maintained the essential farmyard stock, beasts for the plough, with chickens and eggs for rent and consumption. The cat caught the vermin, and to digress slightly, in the same laws, the killer of a king's cat faced a heavy fine, for the animal was held up by the tail with its nose touching a smooth floor, then grain had to be poured over the cat until the tip of the tail was covered. The low angle of rest of dry grain meant that this pile was a very large one, emphasising the value of the cat to the king whose grain it had once protected from rats and mice. This is a good illustration of the damage which could be done by these rodents but also serves to conjure up irresistible images of a splendid Tom and Jerry situation!

Be this as it may, the settlement described was clearly more than an isolated single farmstead but smaller than a village. The nine buildings, each probably a family farmstead, formed a hamlet, with inhabitants often linked by kinship, but here a qualification is necessary for the Welsh law cited can hardly refer to conditions later than the thirteenth century, indeed probably refers to the twelfth century or even earlier. It has no relevance in modern Wales, and probably even in medieval Wales was a stylised model. Nevertheless, the law is important because it suggests that many of the present single farmsteads of that country may once have been small clusters – seen in the presence of the element *-tre* and *-tref* in place-names on the modern map.

The bewildering diversity of terms is part of a rich underlying heritage of language describing settlement types, terms created by the indigenous farming communities rather than scholars, who must now struggle to grasp the inner meanings and subtleties, the similarities and the differences, concealed within the words. Thus, the German *Weiler* derives ultimately from the Latin *villa*, yet another term used to describe a rather large single farmstead, which would have comprised both a great house for the rich owner and dwellings for the farm labourers and slaves who tilled the estate, i.e. would have contained at least the population of the Welsh hamlet described above, perhaps many more, and may have involved several distinct physical entities, the great house itself together with subsidiary hamlets.

Counting farmsteads is one measure of settlement size: in a national map of Great Britain compiled in the early 1960s Thorpe chose to count 'homesteads', which he defined as 'dwellings with dependent buildings and ground', and concluded that a hamlet comprised from three to nineteen of these (Thorpe in Watson and Sissons 1964: 359). He was compelled to adopt this procedure because English maps do not differentiate consistently between farmsteads and other dwellings. However, his figures may not be wholly incompatible with the figure of six to eight farmsteads cited on p. 16, for this contains elements of a traditional historical appraisal and by the nineteenth century many hamlets had been substantially enlarged by the addition of labourers' cottages and other dwelling houses. The German scholar Uhlig adopted a definition for the hamlet (*Weiler*) of between 'three to 10–20 houses and/or farmsteads' which in practice pushes the size into dimensions the present author would feel are more truly villages (Uhlig and Lienau 1972). This is a key problem when dealing with all rural settlements and provides an important focal point for many problems: it is important to remember that in terms of physical size, the village of today was often the hamlet of last century and perhaps the single farmstead of several centuries earlier.

Furthermore, in former centuries the English word 'hamlet' normally carried the idea of dependency, of attachment to a more important focus. Modern British settlement classifications lay more emphasis on function, with hamlets generally lacking such features as post offices, garages and shops, but a key feature is often the absence of a parish church, normally present in a true village. Of course, all criteria are culturally and regionally dependent. In those parts of North America dominated by single farmsteads, the term 'hamlet' is applied to small clusters with the lowest-order central place functions, e.g. a primary school, petrol station, shops, church, etc., which are important for supplying the outlying farmsteads, while the French *hameau* is used for settlements, sometimes quite large ones, which are not the *chef-lieu*, or principal centre of a commune containing the local municipal administration – a usage retaining an older idea of a hamlet being dependent upon another place. Of course, even in regions of Europe where single farmsteads dominate, church-hamlets are still to be found: small clusters of farmsteads and other dwellings set adjacent to the parish church. To conclude, hamlets cannot easily be distinguished from the smallest villages and are identified by three primary characteristics: they are relatively small, they are frequently administratively linked with other places upon which they are dependent, and today they normally contain a more limited range of services than villages. All of these judgements are relative to other settlements within any region.

The best concise, neutral definition of *village* is that of the *Oxford English Dictionary*: 'a clustered assembly of dwelling places'. All of the inhabitants were, in former times, employed in the primary sector, farming and fishing, but the term has now been extended to embrace settlements based upon many types of industry. Once again European languages recognise the distinctive qualities of these places – *village* in French, *Dorf* in German and *pueblo* in Spanish. In some circumstances villages can become very large because of industrialisation or political instability. Indeed through a quite dreadful perversion of language the term is now applied by developers to what should, more properly, be called 'housing estates'. Once again size and function are key criteria. A lower threshold is set by the definitions of hamlets, but today large, very large and even giant villages are known, notably in China, the Mediterranean, parts of Africa and Central Europe and the upper thresholds can be best appreciated by looking at towns.

There is no universally agreed concise definition of a *town* (Carter 1972; Hodges 1982). Usually a town is a relatively large cluster of dwelling places, with buildings and people concentrated into a relatively small area. Because it is never wholly dependent upon farming this morphological distinctiveness extends into many other aspects, particularly the relationships between the people in the town and the local ruler or state, between the settlement and the surrounding area and between the individuals within the community. Further to this, tenurial, legal, social and economic qualities also enter the definition. Towns have long possessed specific qualities which demarcated them sharply from the surrounding countryside. *Stadtluft macht frei*, 'town air enfranchises', says the old German proverb, emphasising the difference between the legally defined personal freedoms of the town and the constraints imposed upon the peasants of the countryside. These latter were often servile, i.e. personally unfree, not slaves but bound to the soil of the estate and compelled to seek their landlord's permission before moving elsewhere or allowing their daughter to marry, because their labour helped cultivate his private farmlands. The rights to engage in manufacture and trade, to leave property by will, to move, to govern and administer and have the freedom of creating a will to leave property to children are all rooted in the town. Historically these factors underlie the more familiar, more modern, lists of service functions, of shops and banks, of hospitals and local authority administrations, definitions which themselves avoid the question of the great amorphous congregations of housing estates and even of slum dwellings – shanty towns – now present around many cities. We have therefore a paradox, that settlements with as many as 8,000 or even 10,000 people, but predominantly engaged in agriculture, such as the 'agro-towns' of southern Italy, are generally considered villages, whereas a settlement with as few as 2,000 inhabitants but engaged in craftwork and trade are urban. In fact, just as there are no clear thresholds between hamlets and villages there are none between villages and towns, and throughout western Europe today many village-sized settlements originated in attempts to establish legal and functional towns. Above all else towns were created to generate and support markets. They were the focal points to which goods could be taken for sale, however this was organised or controlled, and from which goods from the local area could be redistributed and those from further afield spread through regions which could not produce them. Towns and trade are indissolubly linked, and it is this role which must lie at the core of any definition, either historic or modern.

To return, by way of summary, to the key words: a farmstead is an assemblage of agricultural buildings from which the land is worked; a hamlet is a small cluster of farmsteads; a village is a clustered assembly of dwellings and farmsteads, larger than a hamlet but smaller than a town; a town is a relatively large concentration of people possessing rights and skills which separate them from direct food production. These terms conceal as much as they reveal, but they are the building blocks of the study of rural settlement. They express four key ideas and in any country in the world these four elements of settlement may be recognised, although all need not be present within a particular region and varied local definitions must exist. Figure 1.4 incorporates some further ideas concerning definitions, and crudely scales the key words against generalised population figures.

Nucleation and dispersion

As was noted on p. 15, the farmstead is the basic ingredient of rural settlement throughout the world. It is a fundamental building block and can either appear on its own, scattered across the landscape in a dispersed pattern, or concentrated in villages and hamlets which themselves lie spread in a pattern of nucleations.

Figure 2.1 (upper) includes a model of the host of varied possibilities which exist; the circle represents a territory inhabited by a group of farmers, and each sector shows one possible way in which they may locate their steadings within that territory. Each black square represents a single farmstead. There are of course an infinite range of possibilities. The gradation begins with a situation of total nucleation, where the farmsteads may actually touch each other, an arrangement found in nucleations within the Mediterranean and Middle East, through clusters showing varying degrees of separation between individual buildings, to the approximate regularity of complete dispersion. As the dominant nucleation gradually becomes more fragmented – moving clockwise – dispersed entities appear, additional to the nucleation, so that the dominant nucleation eventually becomes indistinguishable in size from other small clusters – hamlets – scattered throughout the territory, although it will often possess a different status. A final stage is represented by complete dispersion, which at its most extreme must involve a wholly regular pattern, to obtain maximal distance between each of the individual elements. This model brings to an uncomfortable focus the realities of settlements seen on the ground, for the diversity seen in this diagram is undoubtedly present. These complexities must be identified, defined, classified and, if possible, explained. The inclusion of two real cases, from Worcestershire, England, shows that in reality even essentially 'dispersed' or 'nucleated' patterns will include mixtures of both nucleated clusters and dispersed isolated farmsteads and dwellings.

Degrees of permanence

Neither settlements nor the buildings of which they are composed last for ever. Müller Wille (cited in Butzer 1964: 340–1) identified a gradation of settlement types based upon

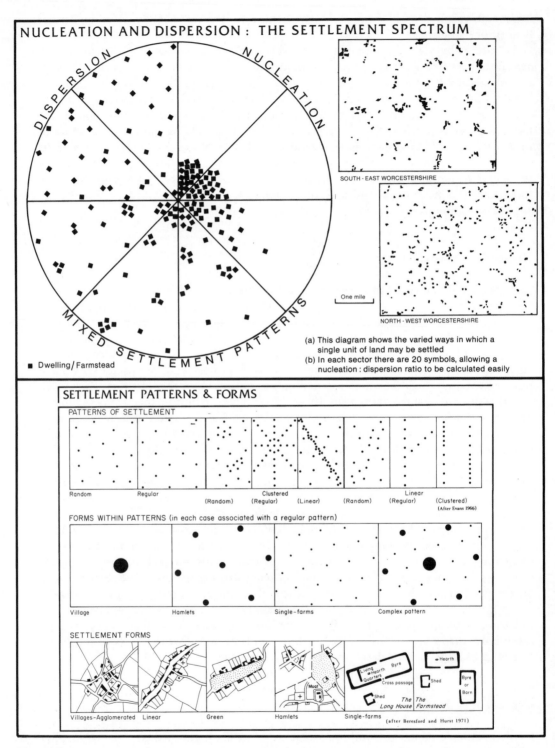

Figure 2.1

the time during which a given site was occupied:

1   *ephemeral* settlements of a few days' duration;
2   *temporary* settlements of several weeks' duration;
3   *seasonal* settlements of some months' duration;
4   *semi-permanent* settlements of some years' duration;
5   *permanent* settlements lasting for several generations.

This list defines a scale: at one end, in ephemeral settlements, the buildings are very simple and the volume of material goods accumulated is very small, while at the other, in permanent settlements, buildings are so constructed as to be very durable – providing they are maintained and not subjected to warfare. They contain the accumulated wealth of human societies rich in material goods, from the personal possessions of individuals to the contents of bank strong-rooms, national museums, archives or art collections. Examples of ephemeral settlements are seen in the camps of Kalahari bushmen made while on the move between seasonal campsites, or the seasonal camps of East African cattle herders such as the Samburu or Masai of Kenya (Wannenburgh *et al.* 1979; Forde 1934: 24–31, 287–307), while of permanent settlements there can be no better examples than the rich urban societies of the late twentieth century. This is an interesting way of classifying settlements, for it concentrates upon neither the forms nor the patterns, merely upon their duration in time.

The permanence of settlement is a function of the power to exploit a restricted environment, and in this the rise of agriculture is crucial. In practice the change from an economy based upon hunting, gathering and fishing to one based upon farming saw an important transition from ephemeral settlements of a few days' or weeks' duration, to temporary settlements for several weeks or months, often closely linked to seasonal activity, or semi-permanent settlements lasting some years, to 'permanent' settlements enduring for several generations. 'What season is this place?' is a question Dodgshon posed to try to catch something of the thoughts of mesolithic, pre-agricultural, people (1987: Chapter 2). The adoption of cultivated crops generated the need to watch and protect cultivated gardens and fields, establishing a vital bond between the farmer and the land in which so much time and labour must be invested. Eventually this helped to bring to an end the shifts to make use of seasonal food surpluses within a group's territory.

By this stage – the neolithic period – there was already a complex interaction between four key factors: the agricultural productivity of the soil, the farming skills of the people, the amount of land or territory a given population could control and use, and finally the arrangement of the population within that territory, whether scattered in dispersed homesteads or concentrated into larger groupings. This is an important point: population density is, theoretically at least, wholly independent of the actual settlement pattern, which may vary between dispersion and nucleation. This can be demonstrated from two examples in Figure 2.2. At the top right the population is concentrated into two clusters – villages – while to the left each family lives in a separate homestead a few hundred metres from its neighbours. This early context, the fifth millennium before Christ, has been chosen deliberately. These first farmers were drawing upon the untapped wealth of soils which had been developing without serious interruption since the disappearance of the ice and the tundra landscapes from all but mountains and the northern fringes of Europe (Champion *et al.* 1984: 125, 131, 133; Piggott 1965: 55). They were establishing

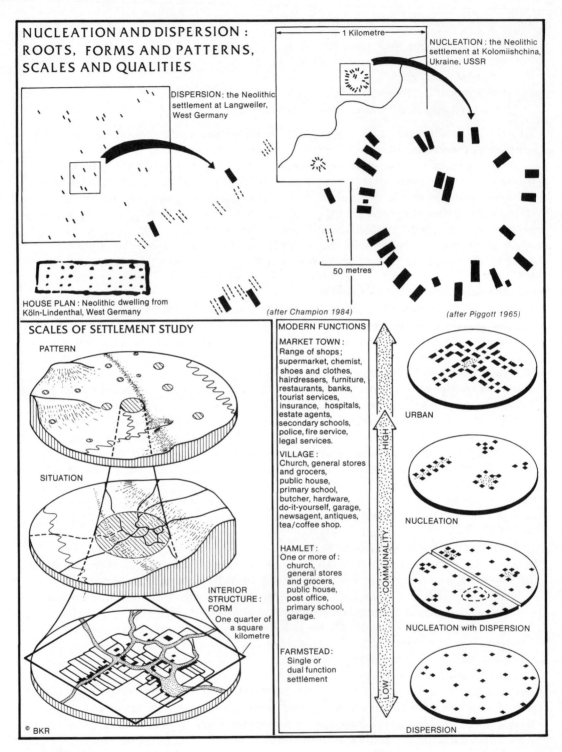

NUCLEATION AND DISPERSION: ROOTS, FORMS AND PATTERNS, SCALES AND QUALITIES

DISPERSION: the Neolithic settlement at Langweiler, West Germany

1 Kilometre

NUCLEATION: the Neolithic settlement at Kolomiishchina, Ukraine, USSR

50 metres

HOUSE PLAN: Neolithic dwelling from Köln-Lindenthal, West Germany

(after Champion 1984)

(after Piggott 1965)

SCALES OF SETTLEMENT STUDY

PATTERN

SITUATION

INTERIOR STRUCTURE: FORM
One quarter of a square kilometre

© BKR

MODERN FUNCTIONS

MARKET TOWN:
Range of shops; supermarket, chemist, shoes and clothes, hairdressers, furniture, restaurants, banks, tourist services, insurance, hospitals, estate agents, secondary schools, police, fire service, legal services.

VILLAGE:
Church, general stores and grocers, public house, primary school, butcher, hardware, do-it-yourself, garage, newsagent, antiques, tea/coffee shop.

HAMLET:
One or more of:
    church,
    general stores
    and grocers,
    public house,
    post office,
    primary school,
    garage.

FARMSTEAD:
Single or dual function settlement

COMMUNALITY   HIGH   LOW

URBAN

NUCLEATION

NUCLEATION with DISPERSION

DISPERSION

Figure 2.2

scattered homesteads, perhaps involving ten people, hamlets of between ten and fifty and villages of between fifty and 250, figures based upon current estimates. To farm at all these immigrant settlers had to clear woodlands (Steensberg 1986: 79–87) but because they were indeed 'the first farmers', the rewards were great and crop yields were, initially, rather high. All subsequent developments were to some degree influenced by the activities of these people, for with the ecologically potent tools of axe, fire and grazing animals they eventually irreversibly altered both large areas of the woodlands of the continent and the soils beneath them. Already, however, the contrast between nucleation and dispersion was present.

To bring the argument to a shorter time perspective, although throughout Europe the rural settlements of today owe most of their immediately visible buildings to the last three centuries, their general layout often took shape in the medieval period (1100–1500). Deeper, earlier, seen for example in the location of many churches, lie roots established in early medieval conditions (400–1100) ... and so on, back to the remoter periods of prehistory. Although difficult to detect, these are often present in such humble features as road lines and the location of key plan elements (Hurst in Roberts and Glasscock 1973: 3–20). Such continuity is nowhere more sharply evident than in Scandinavia, which lacks the discontinuity of Roman occupation, so that prehistory ends with the arrival of Christianity in the eleventh century. In parts of Sweden the pre-Christian barrow burials can still be seen beyond the churchyard wall and a clear distinction between the prehistoric and historic farmsteads is impossible. Furthermore, amid the difficult rock-strewn landscapes of Sweden, the fields cleared by prehistoric farmers are often those of today. In the more favoured environments of western and central Europe, offering a greater range of possibilities for human use, the past is always deeper, more concealed. It is, nevertheless, always present.

In general, crops demand that settlements be fixed, but stock must often be moved to seek grazings. If the economy is a mixed one, involving both crops and stock, then the beasts which winter within the permanent settlement need to walk to more distant locations for summer grazings. So it is that movements of cattle to pastures are documented from many regions of Europe, most characteristically from the Alps and Norway (Braudel 1966: 85–102), while the use of upland shielings was documented in the northern Pennines of England in the sixteenth century (Ramm *et al.* 1970). Such movements took place within countries of the Mediterranean within living memory. In contrast, cultures based almost wholly upon stock must lead nomadic lives moving to follow grazing, water or both. Although now much curtailed by boundaries imposed in the colonial period, seasonal movements remain an integral part of the pastoral economies of Africa, where the occurrence and amount of rainfall are crucial controls over the ways of using the tropical grasslands. The Bahima of north-east Uganda, an upland plateau environment with plenty of grass, are semi-nomadic, moving their kraals every two years, not to seek pasture but to seek water, to try to escape disease and to comply with a custom of abandoning a place where an adult has died and is buried, a telling mixture of practical and psychological needs.

Patterns and forms

The preceding discussion has freely used several terms which must now be defined: two of these, pattern and form, concern the scale of the viewpoint adopted (Figures 1.4 and 2.1), while a second two, dispersion and nucleation, are more broadly descriptive of the character of settlement and may be used in varied contexts. *Pattern* is applied to the distribution of the varied elements of settlement – individual farmsteads, hamlets, villages, market towns or a mixture of all types – throughout a landscape or a region. Single farmsteads may lie wholly separated from each other in a dispersed pattern, or concentrated, with hamlets or villages forming a pattern made up of nucleations. It is possible to talk of a 'dispersed pattern of settlement', a 'nucleated pattern of settlement' or a 'pattern containing a mixture of dispersion and nucleation'. In contrast, *form*, or *morphology*, is applied to the character of an individual element of settlement, be this the form of a house, the form of a village or the form of a town. *Dispersed*, is used to indicate that even the individual farmsteads lie scattered, each being separate from its neighbour, while *nucleated* is used either to indicate a settlement pattern made up of elements in which dwellings and other buildings are concentrated into a number of tight clusters or to describe one such cluster where this is present.

This dual meaning is apt to give rise to confusion, but this can easily be avoided by always qualifying the word 'nucleated': thus a 'nucleated pattern' is wholly different from a 'nucleated form'. Dispersed settlement in rural areas has a pattern where single, solitary dwellings predominate. The houses and buildings are set well apart from each other, whereas nucleated settlement in rural areas has a pattern where buildings are built close together in clusters (i.e. hamlets or villages) and the problem is to determine where dispersed settlement finishes and nucleated settlement begins. In fact 'pure' settlement patterns, wholly nucleated or wholly dispersed, are rarely found; dispersed patterns usually contain at least some concentrations of hamlet or village size, together with some market towns, while total nucleation would mean that everyone lived in one settlement. Like much in rural settlement studies these key words are markers, to be used with an understanding, achieved through experience, of how much they conceal.

To illustrate one problem: how far does a farmstead need to be separated from a village before it is to be seen as a separate settlement unit? Traditionally the 'hailing distance' of 150 metres has been adopted, because beyond this cohesion is lost and neighbours can no longer be called. However, if the peripheral farmer shares the burdens of the village, helps pay the taxes, has arable lands intermixed with those of neighbours, then his farmstead is part of that village community, irrespective of its precise geographical location. Furthermore, participation in the seasonal cycles of activity and bonds of blood and marriage may draw a farmer into the social life of that settlement's community, even if the farm lands are wholly concentrated some distance away around the homestead. Such social and economic factors cannot be ignored, for these bonds can weld even a pattern of widely dispersed settlement elements into a single functioning community.

The lower portion of Figure 2.2 summarises some of the points already made. With the exception of the column concerned with function, which must be altered, even recreated, to fit different times and places, this model is equally applicable in a study of the distribution of the settlements of early farming communities in western Europe in the fifth millennium BC as to the distribution of rural settlements now seen on 1:50,000 maps of Europe or North America. This leads directly, and at times uncomfortably, back to the troubled question of defining the elements of settlement. At a local level this can be made easier by sharpening definitions in two ways: first of all by defining functional variations, and second by thinking not only in terms of buildings but also in terms of the actual levels of population involved (Figure 1.4).

Settlement and population densities

The problem of obtaining even passably accurate figures concerning the actual populations of given rural settlements is profound, even in areas where a formal census is regularly taken. Figures provided by Doxiadis suggest that on a world scale, the gradation from single farmstead to purely agricultural village will normally involve populations ranging between four and 1,500 folk, a generalisation, as the neolithic data cited above show, which is equally applicable in former centuries. Once elements of craftwork or trade are present, then levels of between 250 and 10,000 become feasible, while the village–town threshold is often conventionally placed at about 10,000 (Doxiadis 1968: 65ff.). However, Figure 1.4 emphasises the diversity of views, for even figures derived from near contemporary official sources concerning the village–town threshold show no consistency (UN 1985: x–xiii). In Zaïre, for instance, the urban threshold is to be found in 'agglomerations of 2000 or more inhabitants where the predominant economic activity is of the non-agricultural type and also mixed agglomerations which are considered urban because of their type of economic activity but are actually rural in size', while in India towns are defined as

> places with municipal corporation, municipal area committee, town committee, notified area committee or cantonment board; also, all places having 5000 or more inhabitants, a density of not less than 1000 persons per square mile or 390 per square kilometre, pronounced urban characteristics and at least three fourths of the adult male population employed in pursuits other than agriculture.

Extreme contrasts are seen in Japan, where the urban–rural threshold is set as high as 50,000, while in thinly populated Iceland a town may have as few as 200 souls.

These numerical definitions ignore two further dimensions of the problem; first, the difficulty of identifying the urban–rural threshold when dealing with aggregated populations of large regions, and second, the fact that functional classifications of settlements are particularly time-bound, i.e. the data will relate to only one short time period, although the settlements themselves will normally be very much older, having both added and lost functions through the years of their existence.

Rural and urban balances in England and Wales

In England and Wales Craig (1987) adopted an essentially threefold classification of 'urban', 'mixed' and 'rural' for the populations of wards on the basis of the dominant land-use present (Figure 2.3). Such classifications are never perfect. They are dependent upon the size of each data cell and, no matter how accurate the initial data, they involve elements of compromise, even subjectivity, in defining the categories. However, at a

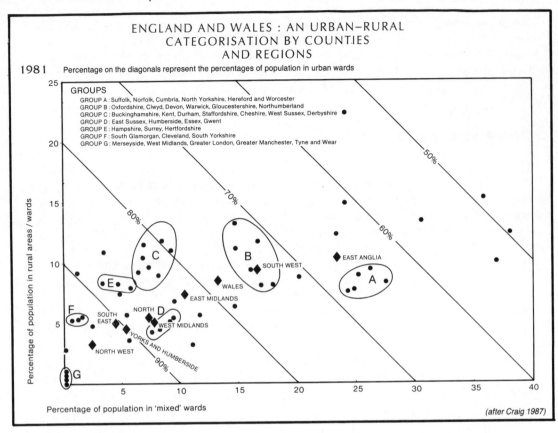

Figure 2.3

national scale and even a county scale, they do permit useful generalisations. The graph incorporates an interesting and ingenious combination of three variables, plotting the percentage of the 1981 population in rural areas against the percentage of the 1981 population in mixed areas, and then scaling these, by means of the diagonals, against the percentages of population in urban wards. The diamonds in the graph represent regional summaries, and taking the case of East Anglia approximately 66 per cent of its population is in urban wards, 23 per cent in rural wards, and 11 per cent in mixed wards, totalling 100 per cent. Table 2.1 summarises the county-scale situation.

Particularly striking is the small percentage of rural population relative to mixed and urban, for the highest rural percentage for a county is in Powys, with about 38 per cent purely rural. In fact the most rural portion of mainland England and Wales (i.e. excluding the Isles of Scilly) is south Herefordshire, where as much as 70 per cent of the population is in this category, while high proportions are also found in parts of Wales, the Eden valley of Cumbria, along the Welsh Borders, in Cornwall and, rather surprisingly, in south Northamptonshire. This list omits that most rural of counties, Northumberland, simply because the figures are all relative to each other, and rural Northumberland has overall a *low* rural population. Such figures are interesting measures of the population living in rural areas, but relating them to the on-ground realities of settlement presents considerable problems, for by their very nature they generalise and obfuscate detail.

Table 2.1 The population composition of counties and regions in England and Wales

Urban population (as % of total)	Rural population relative to mixed	Counties*
64–68	Large	Suffolk, Norfolk, Cumbria, North Yorkshire, Hereford and Worcester
71–75	Average	Oxfordshire, Clwyd, Devon, Warwick, Gloucestershire
80–84	Small	Buckinghamshire, Kent, Durham, Staffordshire, Cheshire, West Sussex, Derbyshire
85–89	Large	East Sussex, Humberside, Essex, Gwent
	Small	Hampshire, Surrey, Hertfordshire
93–94	Very small	South Glamorgan, Cleveland, South Yorkshire
99–100	Nil	Merseyside, West Midlands, Greater London, Greater Manchester, Tyne and Wear

* Listed within each group in ascending order of urban population percentage.

Functional village–hamlet thresholds: an example

To explore an example of the village–hamlet threshold in a specific functional context data for the region around Bury St Edmunds published in 1967 (Everson and FitzGerald 1969) are used in Figure 2.4a to classify settlements on the basis of service provision in 1961. In this case the functional threshold between a village and a hamlet was decided empirically, with the former having a butcher and hardware/appliance shop. This distribution, categorising a hierarchy on the basis of the services present at a given time – for these will now have altered, with post offices, shops, schools and petrol pumps closing as car mobility has increased since 1961 – creates a picture of two contrasting micro-regions. Figure 2.4b plots the same settlements but this time creates a hierarchy on the basis of population, identifying three categories of town, and four categories of village and hamlet. As might be expected, there is a general and close correlation between size of settlement and its level of services, but this was inevitably built into the data actually plotted, for the village–hamlet threshold in terms of services, defined above, does appear at a fall from 624 to 520 inhabitants.

The services present in 1961 have developed within an overall pattern which was inherited from both the recent and the more remote past, for the presence of settlement foci, indeed their very existence, can only be understood in terms of the regions shown in Figure 2.4c. To the south and east is a traditional region of wood pasture, an enclosed landscape, developed on strong, rather heavy, loams of boulder clay country, and three or four centuries earlier a region of dairying and pig-keeping with some horse-breeding (Thirsk 1967: 40–9; see also Figure 3.8). In the north is a zone where shallow wind-blown sands directly overlie the basal chalk, a portion of a larger region known as the Brecklands, which even today are regarded as sub-marginal for cultivation, and are now much afforested. Traditionally this was a region of sheep–corn husbandry, using the power of sheep to graze the fallows and restore some heart to the light soils for further arable cultivation by means of their urine and dung. To the south-west, the exposure of chalk ridges intermixed with valley gravels gives a region of light loams, a much more attractive, productive and variable terrain, again traditionally given over to sheep–corn

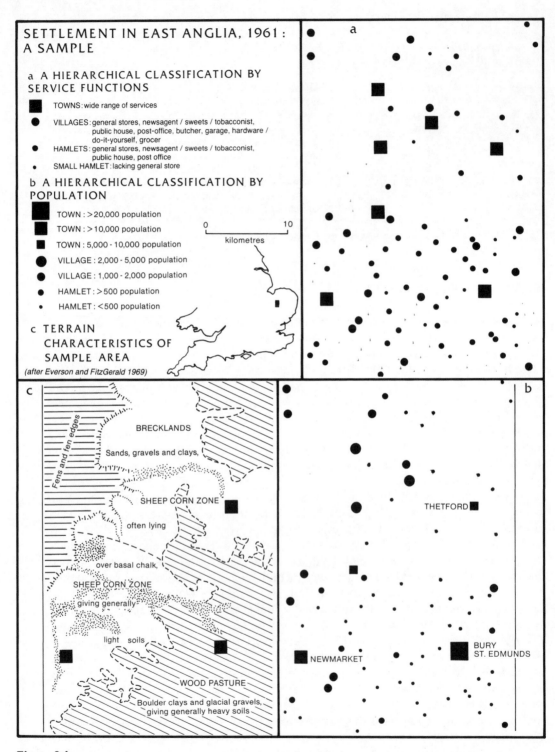

Figure 2.4

husbandry. It was the introduction of turnips into the sheep–corn rotations in such regions which produced vastly greater yields during the eighteenth and nineteenth centuries. Finally, to the north-west lies former fen and fen edge country, with nucleated settlement being restricted to favoured sites amid a landscape that before drainage and reclamation was liable to flood, a zone where stock fattening, horse-breeding and dairying were traditionally linked with fishing and fowling. These ancient antecedent landscape patterns are crucial for understanding the historic forms and patterns of settlement, for wood–pasture regions tend always to be dominated by hamlets and dispersion, the Breckland heaths by almost complete dispersion except for the occasional large nucleation, while sheep–corn country further south supported both villages and hamlets, and a re-examination of the settlement maps will show that regional divisions more subtle than the basic north–south contrast are in fact present, reflections of the historic roots.

Rural settlements and their contexts

Figure 2.5 returns to the detail of a particular place, and is a model of the key structural elements and key relationships found within any purely rural settlement of hamlet, village or even small town size. At the centre lies the settlement itself, in this case a nucleation, but anything from a single farmstead to a market town is equally feasible. Around this lie the arable fields and inner pastures, the meadowland and grazing lands, while further out – at least in earlier centuries – are untamed landscapes, wetlands, heathlands, moorlands, woodlands or mountain wastes. This outer fringe is today more likely to be the territory of neighbouring settlements, indeed their arable fields may well intermingle, with the untamed landscapes being reduced to a few carefully managed fragments. The pattern seen here may be compared with that in Figure 2.6, a model of site and situation which is a diagrammatic but rather fuller representation of *all* the resources conceivably available to wholly rural settlements. It must be stressed that Figure 2.5 depicts one form of land-use zonation and can be elaborated or simplified in many ways: for instance, by varying the nature of the terrain supporting the settlement – as seen in the inset diagrams – or by altering the economy or adding improved communications, a better road system or railway, giving stronger links to other places, both larger and smaller. Above all, cultural and natural landscapes around a settlement will be subjected to progressive changes through time.

Categories of influence

The question 'Why is this region dominated by dispersed settlement and that region by nucleated settlement?' is never answered easily. However, a limited number of factors appear to be important, in particular the circumstances of the physical environment, technological and economic factors, social and demographic conditions and more general historical circumstances (Demangeon in Wagner and Mikesell 1962: 506–16). Such generalisations are, however, so broad as to be virtually meaningless without further definition, but their role in generating the basic contrast between dispersion and nucleation can be shown by an examination of three ideas: first, the extent to which

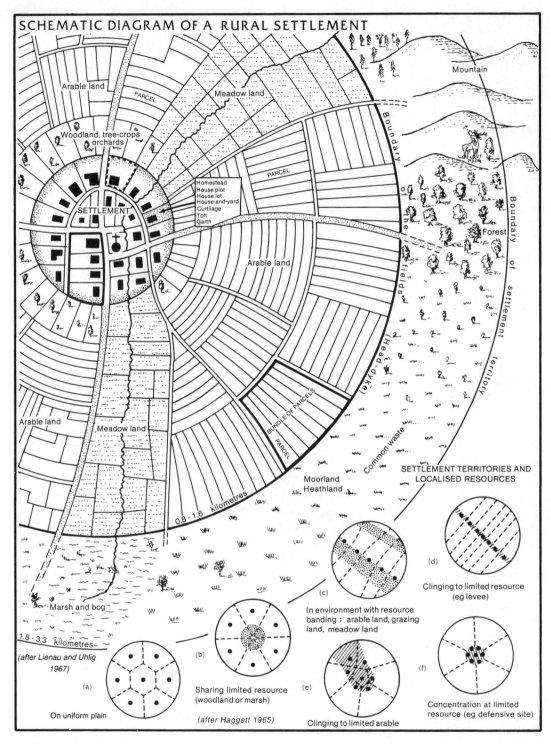

SCHEMATIC DIAGRAM OF A RURAL SETTLEMENT

Arable land

PARCEL

Meadow land

Woodland, tree-crops orchards

Mountain

Homestead
House plot
House lot
House-and-yard
Curtilage
Toft
Garth

SETTLEMENT

PARCEL

Forest

Boundary of the fields

Boundary of settlement territory

Arable land

Head-dyke

BUNDLE OF PARCELS

PARCEL

Arable land

Meadow land

0.8 – 1.6 kilometres

Common waste

Moorland
Heathland

SETTLEMENT TERRITORIES AND
LOCALISED RESOURCES

Marsh and bog

1.8 – 3.3 kilometres

(after Lienau and Uhlig 1967)

(a)

On uniform plain

(b)

Sharing limited resource
(woodland or marsh)

(after Haggett 1965)

(c)

In environment with resource
banding : arable land, grazing
land, meadow land

(d)

Clinging to limited resource
(eg levee)

(e)

Clinging to limited arable

(f)

Concentration at limited
resource (eg defensive site)

Figure 2.5

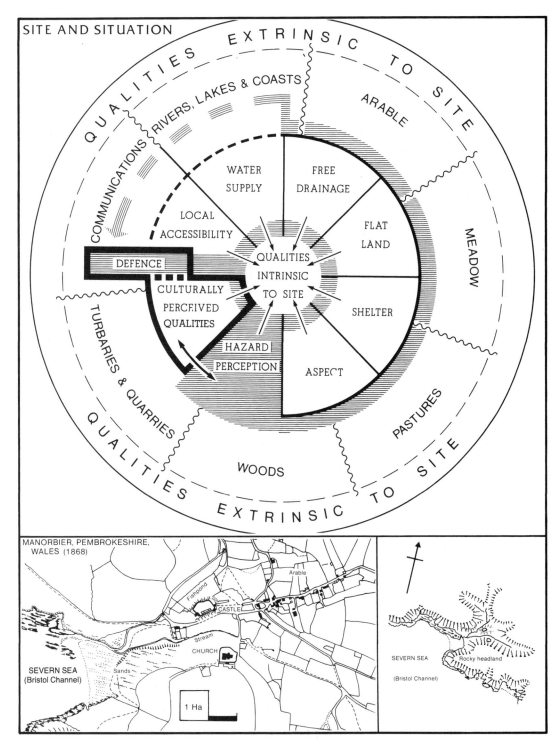

Figure 2.6

physical factors determine the nucleation/dispersion balance; second, the important contrast between site and situation; and third, the underlying social and economic forces which encourage nucleation.

The physical environment

The question of how physical circumstances affect settlement is best approached by building up a simple understanding of the ways in which settlements and the land interact. It has already been emphasised that the relationships between society and the land are never constant or unchanging, for land which is used, however simply, is altered, and successive human occupations must relate not to a pristine, 'natural' landscape, but a landscape already changed by cultural activity, however slight this may be. Furthermore, the presence of 'geographical inertia' constantly brings the past forward into the present; to take a classic example from north-western Europe, many villages have developed on sites which gave access to a water supply, a stream, a spring, a pond or a well, but most villages now draw upon a piped supply system provided by the state or a private company. The original source is now part of a different value system, for the pond, well, stream or spring may now either be rubbish-choked, infilled, buried or piped away, or maintained and beautified as part of a heritage landscape whose qualities are helping to sustain local house prices. This latter was a factor lying far beyond the imagination of the original settlers. In parts of the world lacking piped water, springs, wells, streams and rivers, with all their attendant dangers, are nevertheless still a powerful control over settlement possibilities, for water is needed by both people and domestic stock several times a day.

> If plains are more favourable for villages, then mountains and broken land seem to encourage isolated houses or hamlets. This results from the small size and unequal distribution of arable land, which restrict the efforts of the colonists and prevent them from living together on a single place.

Here Demangeon summarises a common perception of the way in which the configuration of the land affects settlement, explaining the differences by linking together terrain characteristics and soil conditions (Wagner and Mikesell 1962: 506); to these may be added a third important natural condition – water resources. While modern geography tends to be cautious about applying such deterministic explanations, they should not be rejected out of hand. The description of the house in Ghana used in Chapter 1 shows how important they can be. Of course, the physical environment never determines exactly what settlement types are present, but the terrain, climate, soils, vegetation and animal types dominant within an area do influence the lifestyles which can be followed by human societies. The lifestyles which are actually present result from humankind appraising each environment, each landscape, through eyes affected by historical and cultural experiences and in the light of indigenous or acquired technologies.

Thus the Great Plains of the USA offered the native Indians a particularly distinctive buffalo-hunting lifestyle which reached a full flowering only when they obtained feral horses which had escaped from the Spanish *conquistadores*. The first white settlers initially saw these seas of grass as a 'desert', but once the swards were broken by the iron plough in the middle decades of the nineteenth century the Great Plains became the farmlands they now are, with rigidly rectangular boundaries and scattered dispersed homesteads. They can give a rich living, but the delicate soils need careful husbandry and management. The dispersed pattern is the result of settlement by homesteaders, operating individually, and taking advantage of government grants of land (Figure 6.7, eastern Iowa) allocated to colonists as a result of political decisions. Nevertheless, physical conditions are important in that they define and limit the contexts within which cultures generate economic systems and associated settlements. These will be considered in more detail in Chapter 3.

Site analysis

The fundamental linkages between individual places, settlement and the land beneath are to be found in the ideas of *site*, i.e. the area of land upon which a settlement is placed, be this a single farmstead, hamlet, village or town, and *situation*, i.e. the broader locational context of this site, where it is relative to the surrounding terrain. This complex set of relationships is shown diagrammatically in Figure 2.6. Although this has been specifically designed for application within agricultural settlements, whether based upon arable or pastoral agriculture, it is sufficiently broad to take account, for example, of fishing settlements, orientated towards rivers, lakes or coasts, and even mining settlements, where specialist production from quarries or mines is dominant in economic life. In Figure 2.6 the central circle represents the village or hamlet, and this is placed upon an area of land – the site. This may be homogeneous in character or diverse, but it will generally be located within that area of land from which the settlement's population must win a living, its situation. The second circle lists six important qualities which can affect the choice of site – aspect, shelter, flat land, free drainage, water supply and local accessibility. The relative importance of these six will vary from settlement to settlement and from region to region; thus those factors dominant in an area of marshland will differ greatly from those pursued in a mountainous region. The ideas expressed by these key words describing qualities intrinsic to settlement sites remain useful and do form a basic currency of use when considering sites. Nevertheless, if they are applied uncritically it is all too easy to rationalise events through a retrospective, *post hoc* justification. It is inevitable that we seek explanations by looking backwards in time, for it is impossible to escape this viewpoint, but sites were selected not only for their intrinsic qualities but to gain access to the key economic resources which lay outside the site itself, extrinsic qualities, i.e. access to arable and meadow, woodlands and pastures, fuel and building materials, local and extra-regional communications, sea, lake and river. This involved a complex set of evaluations, and each and every settlement site represents a complex interplay of all factors.

It is probable that many of the problems, of water supply, of exposure to winds, of spring or winter wetness, may only appear after a site has been used for several years, when living on the land has altered it, for example by compacting the soil and creating mud in winter or felling nearby timber. An awareness of physical hazards can only result from experience achieved through time, suggesting that some at least of the archaeologically attested movements of settlements may be the result of faulty evaluations during the initial process of gaining experience of landscapes. However, hazard perception is indissolubly linked to the lifestyle, the culture of a given community. Fisherfolk will tolerate very wet conditions with periodic floods if they bring good catches, indeed many even live over, or on, the water. The idea of culturally perceived qualities opens a Pandora's box of questions, for we all see the world and life through the lenses of our own lifestyles, experiences and beliefs. The need to move a settlement when someone dies is not part of modern western culture, but this, and other magico-religious beliefs were part of the everyday lives of all former communities. In westernised societies they are now relegated to the edge of life, the material of curiosity or mere proverb, yet they may well have played an important, if imponderable, part in site selection in former centuries. We cannot recover these thoughts for, as has been noted, 'Dead men do not answer questionnaires.' If a village site has been occupied for more than a few decades, and particularly if it has been occupied for several hundred, we surely delude ourselves if we think we can easily rethink these 'common thoughts about common things' (Maitland 1897: 596). Of course, this should not mean that questions are not asked; common sense has been an important measure of human beings from time immemorial.

The following description of a settlement's site and situation raises particularly interesting questions:

> Only about three miles from Pembroke Castle is the fortified mansion known as Manorbier [Figure 2.6, lower] ... the house stands, visible from afar because of its turrets and crenellations, on the top of a hill which is quite near the sea and which on the western side reaches as far as the harbour. To the north and north-west, just beneath the walls, there is an excellent fishpond, well-constructed and remarkable for its deep waters. On the same side there is a most attractive orchard, shut in between the fishpond and a grove of trees, with a great crag of rock and hazel-nut trees which grow to a great height. At the east end of the fortified promontory, between the castle, if I may call it such, and the church, a stream of water which never fails winds its way along a valley, which is strewn with sand by the strong sea-winds. It runs down from a large lake, and there is a watermill on its bank. To the west it is washed by a winding inlet of the Severn Sea which forms a bay quite near to the castle and yet looks towards the Irish Sea (i.e. westwards). If only the rocky headland to the south bent round northwards a little further, it would make a harbour most convenient for shipping. Boats on their way to Ireland from almost any part of Britain scud by before the east wind, and from this vantage point you can see them brave the ever-changing violence of the winds and the blind fury of the waters. This region is rich in wheat, with fish from the sea and plenty of wine for sale. What is more important than all the rest is that, from its nearness to Ireland, heaven's breath smells so wooingly there.

This graphic account contains more than a touch of estate agent's jargon. This much-loved place, Manorbier in Pembrokeshire, is here described by Giraldus Cambrensis – Gerald of Wales – and was his birthplace and childhood home. Churchman, scholar and traveller, he was born in 1145 or 1146 and died in 1223, having written his book, *The Journey Through Wales*, before 1215. Seven centuries and translation from the original Latin separate us from this tall, sharp and energetic man, with his thick straggling eyebrows – descriptions do survive – but his account retains its magic (Thorpe 1978: 150–1). He was the son of Anghared, granddaughter of Rhys ap Tewdwr, prince of South Wales, and a Norman knight, Gerald of Windsor, castellan of Pembroke; three-quarters Norman, he had a deep pride in his links with an ancient race. The fact that Gerald lists many of the factors appearing in Figure 2.6 provides a measure of support for retaining them as part of attempts to understand settlement. To this day the climate of Pembrokeshire is sufficiently mild to support a commercial vineyard which produces 20,000 bottles per year. Gerald's splendidly biased account contains many value judgements, but even after seven or eight centuries his love of Manorbier shows. It is sobering to reflect how many thousands of such judgements, some rational, others optimistic, made through many centuries of occupation, must underlie a single settlement pattern.

Social conditions: communality

In the light of the preceding discussion, a basic question may be rephrased and focused: what generates nucleations? What forces encourage human beings to congregate into clusters? If dispersion enshrines an idea of individuality, then nucleations derive from a blending of factors to which the term *communality* may be applied. This point demands emphasis; nucleation is undoubtedly linked to the idea of collectivity, in the sense of individuals acting together, but the word 'collectivism' is too closely linked with the ideas deriving from the work of Marx. Accordingly, 'communality' has been adopted to embrace the ideas inherent in the cohesiveness of a community. It covers three interlocking and interdependent ideas, communality of assent, communality to economise and communality of enforcement or coercion.

Communality of assent

At root, human beings evolved in family groups and evolution crossed a critical social threshold when the sick and the old were not merely left to the mercy of predators, but were cared for amid the food-sharing activities around the hearth. In origin we are social animals, and these relationships extend beyond the immediate hearth of one family group to those of adjacent families, related in some way, perhaps sharing a common ancestor. In simple societies, this relationship is often cemented by exchanging women in marriage. The size of the group which could live together under a hunting–gathering economy was conditioned by the carrying capacity of the land, and before farming developed we must picture complex seasonal cycles of activity, splitting bands apart in lean seasons, but with assemblies of larger groups when food was plentiful, for ceremonies and trade. At the root of all nucleations lie family ties, bondings of blood and affection, and these links can be described by the term *communality of assent*. This is a difficult concept for many of us trapped in the later twentieth century to appreciate, for the severance of even nuclear family ties is now commonplace. Not only do family ties remain important in other cultures, attachment to kin was once part of western culture. In a situation in which livings are gained directly from the land, such links interlock with usage rights, landownership and inheritance practices, the crucial questions for any individual being 'To what group do I belong?' and 'What rights do I gain from this?' 'To share in the land and its produce?' 'Within this community?' Of course, communality of assent is by no means invariably linked with nucleation. Consider a landscape of scattered farmsteads, each separated by 300 metres from its neighbours. These farmsteads are by no stretch of the imagination nucleated, but may possess strong ties, coming together to assist kin, or neighbours, in certain labour-consuming tasks when such assistance is essential. Nevertheless, the existence of hamlets and villages sustained by continuous and successful farming or fishing created tremendous opportunities for increased social interaction, choice of marriage partners, ceremonies and merrymaking as well as increased economic potential. Examples may be taken from two harsh environments. In northern Europe during the ninth and tenth centuries Viking settlers, from Norway and Denmark, established colonies throughout the northern seaways, in the Faroes, Orkney, Shetland, some of the Scottish islands, Iceland and Greenland, even creating short-lasting footholds in North America. Initially these new settlements were often the farmsteads of stern individualists, together with their families, kin and household slaves (Figure 1.3). Nevertheless, it is probable that even at the time of foundation there was always a tendency for there to be four adult males in any farmstead or farmstead cluster, sufficient to form a minimum boat's crew. When sons were born, the single farmstead often became a kinship hamlet, two or three dwellings and service buildings. Overlain by many layers of later settlement, these dispersed farmsteads form the foundation of settlement to this day, mixtures of small clusters where arable land is available to support larger numbers, with single farmsteads where resources are more constraining. A very similar physical environment, but colder and less stormy, is to be found in the Åland Islands in the Baltic. Figure 6.5 incorporates (top) a small section of this intricate island archipelago, whose land is still steadily rising after the end of the Ice Age, and maps the farmsteads and other dwellings and arable land. The constraints of the environment are clear: there are pressures towards dispersion occasioned by the fragmentation of good land for arable and meadow, but the difficulties of the same environment demand collaboration by groups of individuals linked by ties of blood and neighbourliness – and the Åland distribution emphasises that in such conditions a mixture of grouping and dispersion arises in similar environmental circumstances but in response to subtle local variations.

Communality to economise

To anticipate a conclusion of this discussion, a village involves more than a kinship grouping; villages are created by pressures which draw or push together people who are not necessarily kin, to create an association whose productive potential far exceeds what could be achieved by each farmer acting as an individual. Agricultural labour, particularly under primitive conditions, is back-breaking: from felling timber and clearing the land, to ploughing, stone-picking, dung-carting, harrowing, sowing, weeding and bird-scaring, to the eventual labour of reaping and carting home the harvest. Each of these words defines a task, the input of many hours of labour to achieve the desired goal. The old proverb 'many hands make light labour' is not true, but many hands do make heavy labour bearable. Villages have been a successful form of settlement because they were successful production units and villages and their associated field systems are bonded together by labour demands. Their success is evident from what they achieved. European culture grew from interaction between the temperate regions of the centre and north and the Mediterranean zone of even older culture to the south. The production of vine and olive, wheat and barley, oats and rye generated surpluses which supported churchmen and knights, merchants and craftsmen, seamen and scholars, and ultimately state and empire. These are themes to be reconsidered in Chapter 7. Communality to economise is seen above all in the by-laws governing the workings of communally organised field systems which for many centuries dominated some of the more productive areas of Europe. There is no doubt that these could stifle individuality, but the fact that they endured shows that there was communal benefit from them.

Communality of enforcement

One of the most distinctive characteristics of the villages of Europe is the extent to which they contain or lie near castles, manor houses or halls and churches, or have in some way been subject to the rule of an aristocratic landowner. In the medieval period, in societies lacking much currency in circulation and banks, people, settled upon land, represented both political and economic power. Peasant agriculture would, at least in good years, generate a surplus of grain and stock, and in western Europe villages seem to have appeared along with feudalism – a society organised so that a mass of peasant farmers supported knights and churchmen. It is possible that most European villages of today occupy sites chosen a thousand years ago, at a time when scattered farmsteads and hamlets were being drawn together. Sometimes this took place on a new site, but often occurred near a long-established hall, for lords needed labour to till their home-farms and sustain revenues. To put a knight upon a good horse or lavishly decorate an altar was an expensive business. Increasingly, this process of concentration or aggregation is seen as an important force creating villages – perhaps (Figures 3.2 and 7.4) it is even the dominant reason for their presence in such numbers throughout Europe.

Thus, the lord's farmstead, and with it a castle, hall or manor house, church and the farmsteads of tenants of varied status, free and unfree, farmer and cottager, are, along with the kinship hamlet, the building blocks of the European village. The farmers worked the fields and paid rents to the lords, who could then fight, pray, build and consume. At core, this process involved bringing families who were not blood relatives into a single community and, often, a single productive unit. In certain circumstances, although the vast majority of villages were not conceived as defensive in a military sense, there was undoubtedly a measure of safety in numbers, while it is possible that some groups of villages were planted as part of a strategic plan to settle friendly colonists. Nevertheless, we must be cautious, for many modern cases of village creation, for instance in Algeria, Vietnam, Ethiopia and Kenya, are as much linked to controlling the inhabitants within the settlement as to defence against threats from outside. Certainly, this factor cannot be excluded, leading to the uncomfortable conclusion that coercion may be at least as important as co-operation in the appearance of nucleated villages.

Conclusion

In this chapter, questions of terminology and definition have loomed large, i.e. the words used to describe, and thus define, the subject matter of settlement study, together with the physiogeographic and anthropogeographic frameworks within which settlements develop. The discussion has not concerned any single time period or locality, but the emphasis has been upon achieving a broad view. To proceed further requires a review of the ways in which the interactions involved in the development of settlement can be categorised.

CHAPTER 3

Explanatory contexts

Theoretical models of settlement must be approached with imagination, for they can only be appreciated and understood if they are keyed into specific examples drawn from reality. The substance of this chapter deals with varied contexts within which settlement forms and patterns can be explained, but the problems involved can be illustrated using three maps of national distributions presented in Figures 3.1, 3.2 and 3.3.

Three national patterns

Figure 3.1 is a remarkable map of Israel. This version is slightly generalised, but it is derived from a study which shows the physical sites of all the villages and hamlets of Israel (Survey of Israel 1970). Like many countries around the Mediterranean the historical need for security in a region often troubled by the passage of armies, coastal piracy and local warfare has stimulated the concentrated occupation of hilltop and spurtop sites. Developments during the last hundred or so years have created a clear division; Arab villages have tended to remain on the older upland sites, while Jewish villages, many of them deliberate plantations, are associated with lower sites, on the plain, in valleys, on low spurs or on alluvial fans, on farmland won by colonisation and irrigation. The result is that the varied types of site occupied reveal something of the Jewish–Arab division which permeates the contemporary state of Israel. This stark and clear-cut break between two cultures is a reflection of recent history, but also serves as a reminder of similar differences, between Muslim and Christian in the Balkans, between German colonists and Ukrainians in the USSR, and between Catholic and Protestant in Germany. Israel provides a closely documented example of the way in which eddies and shallows within the great tidal flows of history affect individual communities, at first separating them, and then isolating them, encouraging the passionate adherence to distinctive cultural qualities. Similar divisions are often to be found among colonists in the New World, so that the Amish of Pennsylvania and other parts of America preserve traditions and lifestyle from two centuries ago, conspicuous for their industry, frugality and old-fashioned dress using hooks and eyes instead of

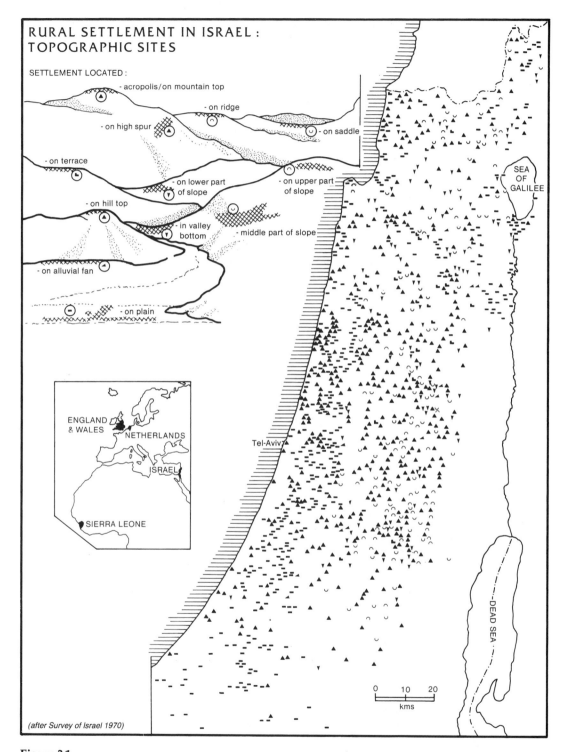

Figure 3.1

buttons. In the Netherlands the single village of Staphorst/Rouveen preserves to this day similar Mennonite traditions, with traditional costumes being worn and traditional practices followed.

The map of England and Wales (Figure 3.2) is based upon mid-nineteenth-century Ordnance Survey maps and shows all nucleations. There are problems with all such synoptic maps of settlement; thus the original sources range in date between 1813 and 1869, the cartographic styles vary from area to area, mistakes could have been made in compiling this distribution. Nevertheless, the 7,500 or so symbols are more than a sample, they are sufficient to reveal the reality of the mid-nineteenth-century distribution (Harley 1964a). At this scale some of the broad responses of settlement to terrain contrasts are crystal clear, and this is a distribution to be studied alongside both a relief map and a soil map. Remembering that this is a map of nucleations, i.e. hamlets, villages and towns, several points are evident: the avoidance of upland areas with limited arable potential (in the Pennines, Wales, Exmoor, Dartmoor) and lowland areas liable to flood (the Fens and the Somerset Levels); the linear patterns in Lincolnshire and Wiltshire, in response to the alignment of vales and escarpments and the presence of chalk where water is scarce. Thus, while in detail many English settlements are sited to give access to water, and the chalkland valley villages are a prime case, this was never the dominant control as in less well-watered countries. There is also a remarkably dense scatter of nucleations throughout the eastern Midlands, reflecting the general and uniformly good potential of the arable lands.

All this is clear, but the white areas raise important questions. There are indeed parts of Wales and the Pennines where areas of as much as 100 square kilometres are virtually devoid of people, but in general the uplands carry a pattern of dispersed settlements. However, essentially the same arguments apply in eastern England, traditionally the most populous and agriculturally prosperous part of the country, which was in the mid-nineteenth century dominated by dispersion. Furthermore, across the Midlands, the nucleation/dispersion break, so clear on the map, accords with no physical boundary on the ground; it is a cultural feature. Of course, these arguments are framed so as to generalise, and this complex map is introduced here to provide a framework of reference. The broad lesson is clear; physical factors can be important in understanding the nucleation/dispersion balance at a national scale, but they are never free of other, often more powerful, cultural influences, deriving from the character of human societies.

Can this be tested elsewhere? Sierra Leone provides a splendid opportunity, for David Siddle has gone to the trouble of creating a real rather than a generalised distribution (in Clarke 1966: 60–1). In this case (Figure 3.3) the map is based upon topographic maps and air photographs, and is considered to be '80 per cent accurate'. Looking carefully, four principal zones may be distinguished: areas wholly devoid of buildings; areas of generally low settlement density, forming a great wedge in the north and east of the country; a southern area where a dense concentration of large nucleations is enhanced by pockets where great numbers of small clusters are also present; a northern and western zone, with patchy concentrations and very mixed character. The empty lands and areas of low settlement density indisputably relate to the presence of coastal mangrove swamps, forest reserves and higher hill masses, but the north and west, a dissected plateau and often difficult of access, has had a long history of inter-tribal warfare. In general the areas of high settlement density are explicable in terms of agricultural productivity and the presence of cash crops, such as swamp rice in the

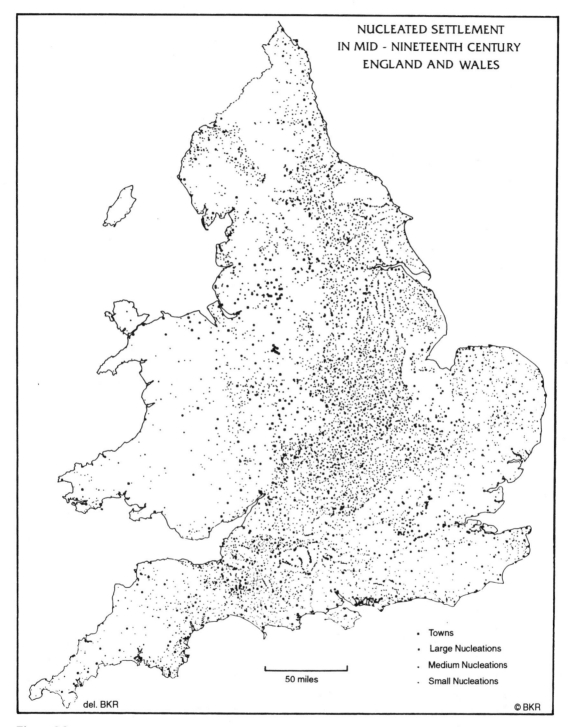

NUCLEATED SETTLEMENT
IN MID - NINETEENTH CENTURY
ENGLAND AND WALES

• Towns
• Large Nucleations
• Medium Nucleations
• Small Nucleations

50 miles

del. BKR

© BKR

Figure 3.2

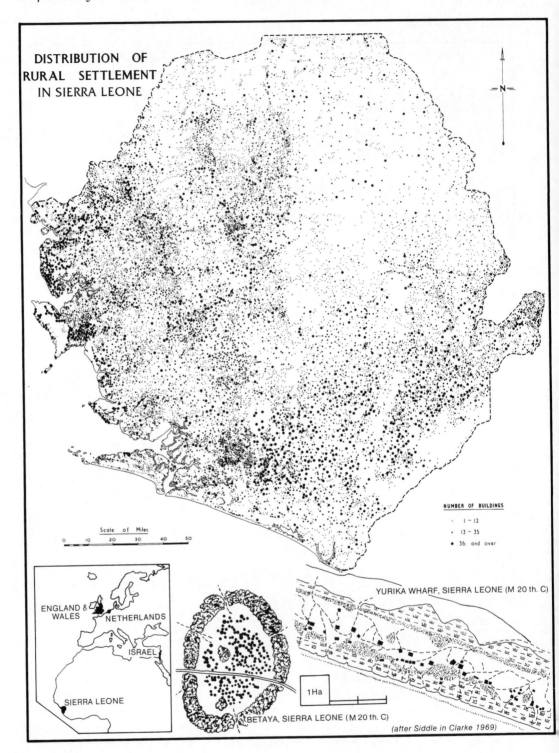

Figure 3.3

north-west and tree crops in the south-east, but it is equally clear that Siddle, knowing the ground, had difficulty in explaining his fine map. Land tenure, recent colonial history and the type of agricultural economy practised have all played a part. Thus the distinctive southern and eastern zone of rather larger nucleations corresponds almost exactly with the presence of the Mende people of the Upper Moa Basin, so that here the larger settlements may be a cultural trait, although it is evident from other maps that this is also an area with the greatest level of reliability of rainfall in the dry season and conversely the lowest percentage of its annual rainfall in the wet season. A reliable water supply may be a crucial factor permitting the presence of larger nucleations. Significantly, this same area has the greatest number of towns, most thickly concentrated along the north-western edge of the Mende territory. This hints that the 'empty quarter' was at some stage a procurement zone, exploited from the more prosperous area to the south, and that these frontier towns would be markets for the exchange of slaves and other products.

None of these analyses is in any way complete, indeed each of these maps is a research tool: all three represent, in Siddle's words, sources 'from which a great deal of information has yet to be culled'. This is as true of Israel and England and Wales as it is of Sierra Leone and this is at once disturbing and exciting. It is disturbing, for there are few areas of the world as closely studied as England and Wales, and we would expect to be able to answer a question as simple as 'Why do we get nucleation here and dispersion there?' This is not yet possible, any more than sweeping explanations can be offered for the settlement distributions seen in Sierra Leone. It is exciting, because questions are always more exciting than answers, for they represent challenges. What is certain is that in none of these cases is there a simple land/settlement relationship wholly determined by physical circumstances.

These three national distributions give a feeling of the sheer scale and diversity of the earth's settlement mantle and serve to emphasise the diversity of factors which must be taken into account within explanations. Two fundamental approaches must now be assessed; first, given that physical factors are undoubtedly of prime importance in framing the general presence or absence of settlement in three contrasting environments, sometimes attracting, sometimes repelling, their role must be considered in rather more detail. Second, as a way of generalising about the specific approaches adopted by scholars attempting to explain particular circumstances certain core concepts and ideas about the way space is arranged and organised must be present.

The role of the physical environment

In practical terms physical geographers divide the world into a series of physiographic systems dominated by distinctive processes, fluviatile, glacial, periglacial, desert, volcanic, coastal, which are further sub-divided on the basis of the climatic zone within which they operate. Inevitably these systems have broad relationships to world climatic zones identified by climatologists, and world zones of vegetation and soils have a relationship to terrain and climate (Mitchell 1973). Although much modified by human agency, these world zones – after all, merely a form of classification – do nevertheless exist and a knowledge of these broad frameworks is an essential foundation. It is probable that ever since conscious thought appeared human beings have always speculated about their relationship with the earth, practical questions which share roots with myth and religion. In a historical analysis of ideas present in geographical writings from ancient times to the eighteenth century Clarence Glacken (1967: Introduction) encapsulated the many questions, ideas, speculations and beliefs under three heads: first, the idea of a designed earth, created as a home for men and women, seen in the myth of the garden of Eden but also found amid the mythic store of many human societies; second, the idea of the influence of the environment on human beings, an idea as Glacken says 'deeply involved with interpreting the endlessly fascinating array of human differences' (1967: 709); third, there is the idea of humankind as a modifier of the environment, most beautifully summarised in a quotation from the Roman writer Cicero:

> We enjoy the fruits of the plains and of the mountains, the rivers and the lakes are ours, we plant trees, we fertilize the soil by irrigation, we confine the rivers and straighten or divert their courses. In fine, by means of our hands we essay to create as it were a second world within the world of nature.
>
> (quoted in Glacken 1967: xxix)

These fundamental ideas are in no way mutually exclusive although concepts of 'environmental influence' and 'man's role in changing the face of the earth' permeate geographical writings. Works by many scholars have revealed the vast pitfalls present in uncritical correlations between physical conditions and settlement. To take one case from England, numerous studies exist which relate the distribution of Anglo-Saxon and Scandinavian nucleated hamlets and villages to drift geology or soil maps, but recent work is showing that the hamlets and villages now seen on the map may in fact be late arrivals, resulting from aggregation during the centuries between perhaps 1000 and 1200 and thus post-dating the Anglo-Saxon and Scandinavian periods. Of course, this then raises the far more interesting question about the settlement system which preceded that based upon hamlets and villages. While excavations of settlements – villages, hamlets and farmsteads – have taken place, it is difficult to reconstruct the pre-1066 settlement pattern for large areas; the evidence is simply not available. Further, the survival of an Anglo-Saxon or Scandinavian place-name emerges as a crucial research question leading to an assessment of the extent to which the name refers to a specific *place* or to a *territory*.

Must earlier studies then be dismissed out of hand? No, because scale comes to the rescue. The physical maps are themselves of variable accuracy, particularly where soils or drift deposits are involved, and there are situations in which earlier communities – and indeed the present farmer – may know much more about local variations than a modern investigator. Even if the crude data do not allow a real discussion of the sites of settlement which precede those visible on the modern map, then plotting a large area on a relatively small-scale map may allow broad correlations to emerge, by revealing, for instance, that settlement may have been attracted by certain soils for agricultural purposes rather than as the sites for specific settlement location. The fact that so many English villages bear Anglo-Saxon or Scandinavian names is a pointer to the presence of some settlement at that stage, even if we know little about it. An understanding of scale is crucial when generalising about settlement, and what we know of forms and/or patterns must always be carefully differentiated and the links between settlement and physical geography must be approached with great caution.

Physical contexts

Much has been written about the relationships between settlement and physical circumstances. If we admit that the interactions between human cultures, their lifestyles, economies and societies are too complex to be susceptible to simple explanation, how then can we progress? A digression may assist. In the ancient stories of the primeval peoples of Ireland, one of the oldest literatures of Europe, a people called Nemed are said to have 'cleared twelve plains and built two royal forts'. More specifically at Tara, the ancient capital of Ireland, this group 'cut down its tangled wood so that *its corn was rich corn*' (Rees and Rees 1961: 110–11). This ancient story – perhaps having ultimate roots in the prehistoric past – provides an imaginative spark. If we could go back to a time when human culture had but little impact upon the environment, how would landscapes have actually been seen by human groups? Not *what* landscapes, but *how* would the landscapes which then existed have been perceived by earlier societies?

As early as 3000 BC Sumerian surveyors were mapping the differences between plain and mountain – drawing their maps on clay tablets – for each offered different ways of earning a living and provided varied products for human use (Steensberg 1989: 35). It is clear that the perception of an upland–lowland contrast was fundamental, and on the plains these people were already measuring and recording land divisions in great detail.

For the moment let us forget the intricate details of geology, drift and solid, or soil and vegetation maps provided by the work of later scholars, and think of the land beneath. The earth's surface is highly varied. Furthermore, the surface alters seasonally, it is never constant, and this is why an understanding of climate is important; but two qualities give it its primary characteristics – relief and vegetation. These are of course closely linked to the stability, or otherwise, of climate and the geology and geomorphological history of a given area. Human beings will always perceive relief – plain, hill and mountain – and note that regions may be well-watered, adequately watered or seasonally or wholly dry, covered with trees or lacking trees. Elemental as this may seem, for it is not disputed that there is a world of difference between a mangrove swamp and a coniferous forest, it allows two factors to be identified: first, at their most extreme these terrain contrasts do undoubtedly affect human activities, not least they provide varied natural resources and settings within which lifestyles evolve. The three words, 'plain', 'hill' and 'mountain', are, of course, vast world-scale generalisations, for there are not only many types of plain and zones intermediate between plain and hill, and hill and mountain, but their psychological importance in earth-culture relationships should be appreciated.

Next we must recognise that all human activity changes the pristine untouched landscape. If one took three plains – the twelve of Ireland complicate the argument too much – and cultivated them, the subtle differences initially present and not necessarily immediately visible would emerge. These would be the result of variations in latitude and location, height and relief, climate and underlying geology, and would eventually become exaggerated by human activity, one plain suffering soil erosion because of over-use, one plain dying as salt water invaded, because cultivation affects water tables, while one plain, deeply favoured, remains 'rich in corn'. Several thousand years later, the visible landscape of each undoubtedly retains many physical qualities wholly derived from the natural world, but each has been culturally altered, and in detail even the soils or stream lines may be as much culturally created as natural. Thus, the story of the creation of twelve plains, mythical or otherwise, is powerful recognition of the ability of human cultures to alter pristine nature, and from such simple beginnings, changes brought about by fire and axe, human culture moves forward to the alteration of both the atmosphere and the stratosphere with by-products of industrial activity.

The conclusion to be drawn from this argument is an important one: the landscapes we now see, involving an assemblage of physical and cultural features which give character to places, are both essential geographical frameworks within which to study all settlement, and in themselves a summation of all the processes, some physical and some cultural, which have created them, through both geological and cultural time. Morgan (1983), writing on Nigeria, uses the neat expression 'achieved landscapes' to convey this idea. We must never assume that landscapes have always been as they can now be seen and experienced, for in past millennia the plains and the mountains may have been unattractive to cultivators who preferred the hills between, hills now wholly bare because of millennia of felling, burning, cultivating and grazing. So often in settlement geography elements of chicken and egg arguments are involved. Which came first? Is that plain a successful agricultural zone because that was what it was seen to be suited for, or did the local hills provide a base from which to colonise an initially unattractive plain, which became attractive only when skills and technology improved, and it was appreciated that the silt-laden flow of its rivers were able to renew and sustain soils without creating excessive erosion from the deforested and cropped surfaces? Farmers, and we must concentrate on farmers, have long been able to recognise not only the larger-scale differences, inherent in mountains, hills and plains, which may affect the ways that whole communities behave – thus possession of a fertile plain may be worth fighting over – but have also accumulated experience of local regions and even the characteristics of individual fields. These are experienced, quite literally, through the soles of the feet during the ploughing process and seen in the quality and quantity of each field's life-sustaining yields. It cannot be sufficiently emphasised that farmers were doing this long, long before scholars became interested in such differences and qualities.

The discussion so far has suggested that there could be only one response to a given type of environment, for the words chosen imply that this response is economic. However, economic life is one essential characteristic of human culture. Within us all lie a curiosity and deep patternings which ensure that we are concerned with more than eating, drinking, sleeping and sex, important, indeed wholly essential, as all these may be for sustaining us as individuals and as a species. As was argued earlier, landscapes are always more than vehicles for economic activity; they may be loved and respected, sought and avoided, feared and hated. It is no accident that some of the earliest monumental human creations were temples, rising stage by stage above the level of the plain of Mesopotamia whose soils and water sustained the farmers whose tribute supported the priests, craftsmen and rulers. The mountain (and of course the temple) 'towered above the peopled plains; it was difficult to approach, dangerous and unassimilable to the workaday needs', uniting earth and heaven (Tuan 1974: 40). Among 'primitive' peoples landscapes normally possess a cosmic significance, seen as inter-mediaries between the practical world and other worlds, transcending the humdrum routine of economic necessities. They have magico-religious properties, qualities we, in the pragmatic cultures of the developed world, are struggling to sustain. Seeing is as much part of psychology and belief as it is of the mechanics and optics of the eye and brain: how we see landscapes is very much part of this and not just a matter of 'what is good for agriculture'. Sacred places, from Ayres Rock to the Mesopotamian ziggurat,

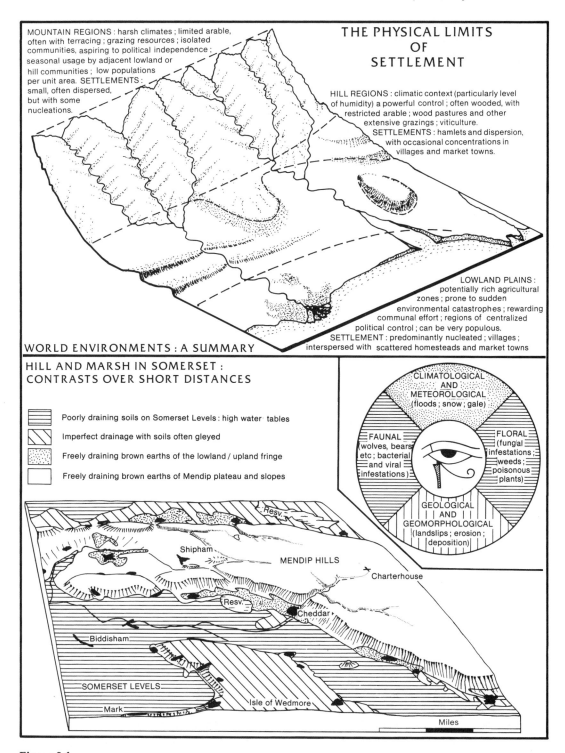

MOUNTAIN REGIONS : harsh climates ; limited arable, often with terracing ; grazing resources ; isolated communities, aspiring to political independence ; seasonal usage by adjacent lowland or hill communities ; low populations per unit area. SETTLEMENTS : small, often dispersed, but with some nucleations.

THE PHYSICAL LIMITS
OF
SETTLEMENT

HILL REGIONS : climatic context (particularly level of humidity) a powerful control ; often wooded, with restricted arable ; wood pastures and other extensive grazings ; viticulture. SETTLEMENTS : hamlets and dispersion, with occasional concentrations in villages and market towns.

LOWLAND PLAINS : potentially rich agricultural zones ; prone to sudden environmental catastrophes ; rewarding communal effort ; regions of centralized political control ; can be very populous. SETTLEMENT : predominantly nucleated ; villages ; interspersed with scattered homesteads and market towns

WORLD ENVIRONMENTS : A SUMMARY

HILL AND MARSH IN SOMERSET :
CONTRASTS OVER SHORT DISTANCES

Poorly draining soils on Somerset Levels : high water tables

Imperfect drainage with soils often gleyed

Freely draining brown earths of the lowland / upland fringe

Freely draining brown earths of Mendip plateau and slopes

CLIMATOLOGICAL AND METEOROLOGICAL (floods ; snow ; gale)

FAUNAL (wolves, bears etc ; bacterial and viral infestations)

FLORAL (fungial infestations ; weeds ; poisonous plants)

GEOLOGICAL AND GEOMORPHOLOGICAL (landslips ; erosion ; deposition)

Resv.

Shipham

MENDIP HILLS

Charterhouse

Resv.

Cheddar

Biddisham

SOMERSET LEVELS

Mark

Isle of Wedmore

Miles

Figure 3.4

from Stonehenge to Chartres and Compostella and the Kaaba of Mecca, are a part of human experience and parts of culturally perceived and culturally moulded landscapes.

To conclude, understanding the physical controls over settlement is dependent upon the scale of study and in Figure 3.4 the upper section portrays a generalised physical landscape, embracing a block of land several hundred square miles in area, while the lower section of the diagram illustrates a block approximately 15 square kilometres in Somerset. At both scales it is often the junctions between hills and plains which emerge as zones attractive to settlement because of the environmental contrasts present. The nucleated settlements of the small area of Somerset are concentrated in a 'preferred settlement zone' on the freely draining soils appearing on the waning slopes between uplands and marsh, sites which also allow easy access to other landscape types. The Anglo-Saxon names of the settlements are one pointer to the period during which they originated, indeed excavation has shown that Cheddar was at that time the site of a great timbered royal palace. Nevertheless, the settlements recorded in the diagram are merely the survivors of a succession of occupational sequences within the region, for long-deserted prehistoric and Romano-British settlements are to be found both on the dry limestone uplands of the Mendips and amid the squelchy marshlands, now drained, of the Somerset Levels.

Conceptual contexts: the ordering of space

The imposition of order upon chaos, thought patterns upon reality, is an important part of enquiry, and such explanations of rural settlement as have been achieved derive from this structuring of observations. While the next section deals with several interlocking ideas concerning settlements, core–periphery concepts, diffusion models, pattern characteristics, and hierarchical models and dependency models, it is in practice difficult to separate these completely from each other. All touch upon different aspects of a total and complex reality.

Core–periphery concepts

Core–periphery or centre–periphery concepts originate in geopolitics and economics and can be applied at many scales, from world and continental geopolitics to purely local circumstances. The root of the argument is that a core region develops as a result of inherent advantages of physical conditions and location, normally the presence of good agricultural resources, abundant minerals available for extraction, fine port facilities or even proximity to a rich neighbour; these singly, or more usually in combination, create a potential for economic growth which is higher than that of surrounding regions. This core becomes the site for the highest and most efficient use of resources, with the ability to attract more in the form of labour, entrepreneurship, unprocessed raw materials and capital. These feed into the core, at first from the surrounding areas but eventually from distant regions. Growth in this core is such that it benignly acts as a centre for cultural and political life, but often, as communications strengthen, it parasitically drains the peripheral regions of resources and people. The periphery becomes the source of raw materials and of demand for manufactured goods produced in the core. It will be clear from this that at a world scale the imbalances between the industrialised nations, western Europe, the USA and Japan, and the Third World are classic examples. During the mid-nineteenth century Britain occupied a key world position in this respect (Figure 7.3).

It will also be apparent that core–periphery relationships are by no means static. Figure 3.5 presents two cases: the upper example, Europe, suggests that many countries of this continent represent accretions to ancient core territories, focal well-populated areas capable of generating a surplus income above the subsistence level. Able to equip and sustain armies, these cores played crucial roles in power politics. From the states developing round them expansion took place into both the peripheral lands nominally under their control – an internal frontier – and through the conquest of other regions – an external frontier (Pounds and Ball 1964: 24–40). This model provides a useful broad view of European history, a framework within which to see, for example, the relations between the core region of the English state based on the scarps and vales of south-east England and the Celtic regions to the west and north, Cornwall, Wales, north-west England, Scotland and, not least, Ireland. The second map, also at a continental scale, shows the chronology of the accretion of territories to the United States of America, a pattern closely tied to the westward movement of the frontier (Billington 1960: 1–11; Brown 1948: 89–110; Blume 1975: 113). This is generally reckoned to contain four elements: the movement west of, first, fur-traders, then cattlemen, followed by miners, and then, finally, farmers. Each evaluated and utilised the environment differently – the stuff of many 'westerns' – but it was the farmers who, in Billington's words, 'made no compromise with the environment: their task was not to adapt but to conquer' (Billington 1960: 5). This process of 'sequent occupance', visible at the broad scale, but appearing again and again at the most detailed levels in each of the varied environments, was repeated for three centuries as Americans colonised the continent. The limited colonies of the eastern seaboard, established before 1750, were followed after the War of Independence of 1775–81 by a nationally pursued policy of expansion west of the Mississippi, while the substance of the present state was defined by 1850 by large acquisitions from New Spain, i.e. Mexico. The concept of the 'frontier', first clearly enunciated in the United States by Frederick Jackson Turner in 1920, can be used in many other contexts. It is in effect one model of the process of diffusion.

The conquest of much of western Europe by Rome is an ancient example (Figure 3.6, lower): an inner core, Rome and Italy, was a consumer of raw materials, manpower and cash in enormous quantities, far above its own productive capacity, and maintained a high level of urban population and the civil and military apparatus of the state. Around this lay an inner periphery of rich provinces, Spain, Gaul (France), Egypt, North Africa, Asia Minor (Anatolia), producing, under Roman management and tax demands, far more than local needs, rendering tribute and trade goods to the core. Beyond this lay an outer periphery comprising remote governed provinces, more or less Romanised client kingdoms, and wholly unconquered barbarians. Within these peripheries were the military zones, sometimes with fixed, sometimes with fluid frontiers, but invariably rather unstable. The local rulers of the barbarian territories adjacent to the actual frontiers were often in receipt of prestige goods, 'gifts' and honorary Roman titles, but these were in effect procurement zones, rendering to the core, Rome, supplies of raw materials, not least slaves, a form of cheap labour.

It is fair to ask why this core-periphery concept is so useful a tool in understanding settlement, indeed many aspects of human geography. Core–periphery concepts have

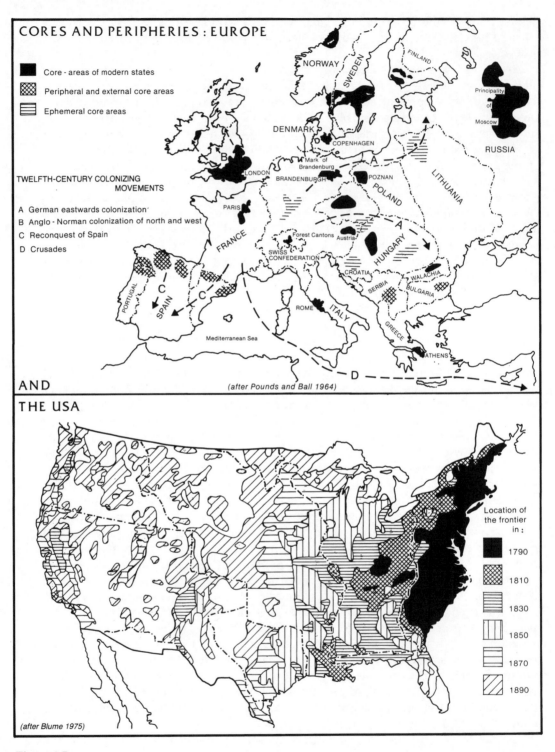

Figure 3.5

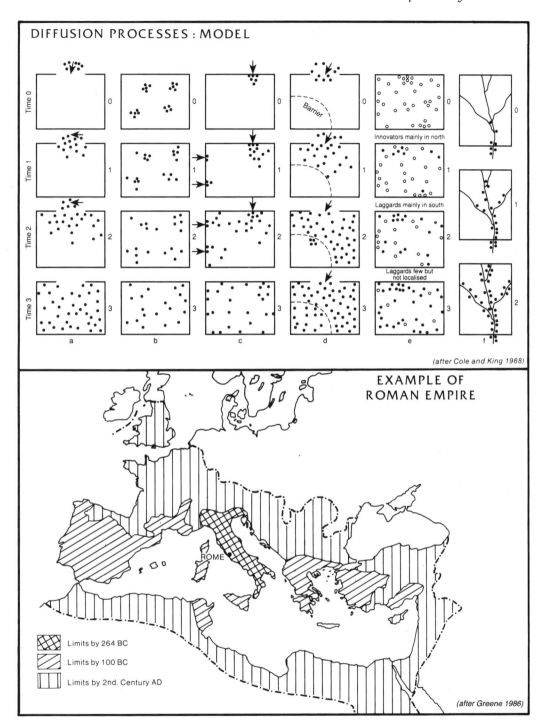

DIFFUSION PROCESSES : MODEL

Time 0
Time 1
Time 2
Time 3

Barrier

Innovators mainly in north

Laggards mainly in south

Laggards few but not localised

a   b   c   d   e   f

(after Cole and King 1968)

EXAMPLE OF
ROMAN EMPIRE

ROME

Limits by 264 BC

Limits by 100 BC

Limits by 2nd. Century AD

(after Greene 1986)

GEOGRAPHY AND TIME : CASES OF DIFFUSION

Figure 3.6

two levels of significance: first, they can identify ephemeral situations during the development of any area's settlement, and the peripheries, often frontiers or border zones, will normally have settlements with different characteristics to those already well established in the core or heartland regions. Second, they can also be used to identify deep and stable structures within a country's developed settlement characteristics; for example, nothing will give the settlements of Scotland or Wales the characteristics of the 50 miles around London. The differences are more than mere physical geography.

The best general answer to what forces are involved in core–periphery developments has been provided by Barry Garner (Chorley and Haggett 1967: 304–5) who defined six fundamental premises concerning places (locations), which show what centrality really represents, and which can then be translated into various contexts, both contemporary and historic. The premises are:

1 The spatial distribution of human activity reflects an ordered adjustment to the factor of distance. Distance is basic to geography because human activities must be scattered over the earth's surface; it cannot be concentrated at one point. Spatial variation and areal differentiation are fundamental to the discipline.

2 Locational decisions are taken, in general, so as to minimise the frictional effects of distance. Movement costs human beings effort. For instance, settlements are not normally sited at any great distance from the arable land supporting them.

3 All locations are endowed with a degree of accessibility but some locations are more accessible than others. Accessibility implies the 'ease of getting to a place', but once again this is never constant; the building of a new road can generate accessibility where none existed before. A common problem is that landsmen will always tend to see the sea as a barrier; it is not, it is a great highway.

4 There is a tendency for human activities to agglomerate to take advantages of scale economies. By scale economies is meant the saving of costs of operation made possible by concentrating activities at common locations. A nucleated community, sharing the agricultural work, can prove a more effective way of sustaining output. Craftsmen, serving a nucleated community, as their skill increases, gain advantage from concentrating where there is a larger market.

5 The organisation of human activity is essentially hierarchical in character. This important point is discussed below, but there are powerful links between social evolution and spatial order (Figure 7.1), and the more accessible locations tend to generate the largest and most successful settlements. In historic terms such locations may be sustained through many phases of occupation; Everitt has called these 'seminal places', seen as important for many centuries and to many cultures.

6 Human occupance is focal in character. This notion is fundamental to the core–periphery concept; some regions, and not just individual places or localities within a region, emerge as nodes, important on a very broad scale. Not only is there a hierarchy of places, it is possible to identify a hierarchy of regions.

To these premises may be added a seventh:

7 Human occupance has a tendency to inertia. The reason for this is part economic, for moving the settlement at the focus of a regional web of activity is an expensive business

– although it can be done, think of new capitals such as Brasilia – but it is also part psychological. Thus York remains an important town in northern England because an Anglo-Saxon ruler consciously adopted it as a capital because he was well aware of its importance in Roman times.

Core regions represent a convergence of the tendencies represented by these premises. In particular they possess accessible locations, experience the agglomeration of productive forces and the appearance of high-status places within an overall hierarchy, and through time see the emergence of a powerful nodality dependent upon routes, together with a powerful inertia, generating new developments within the old core.

Diffusion models

It is obvious that few distributions achieved their present character instantaneously. Diffusion, involving the spread through space and time of defined features, or contraction, the reverse, are likely to be key mechanisms involved in change; indeed diffusion is fundamental to the understanding of core–periphery concepts. Diffusion or contraction may take place rapidly or slowly, may involve physical objects such as people, animals or plants, or concepts and ideas such as technologies or political opinions, and the underlying mechanisms may be spontaneous through vectors, as with a disease, or consciously directed, as in an advertising or propaganda campaign. The whole process is complicated by the fact that diffusion may meet many barriers, both physical such as mountains, or cultural such as religious beliefs, and these barriers may be absolute, or merely act as a sieve, slowing or redirecting the movements. The upper portion of Figure 3.6 models various types of diffusion (Cole and King 1968: Figure 10.17): (a) is a situation in which arrivals from outside enter at a point, and then during three time phases – Time 1, Time 2 and Time 3 – diffuse throughout a defined territory; a good example of this would be the arrivals of Norman invaders in Britain after their victory at the Battle of Hastings in 1066. Column (b) shows a situation in which elements already present within a given territory simultaneously diffuse outwards, while (c) shows a circumstance in which diffusion starts from one point, and then, at a later stage, is introduced to other centres and diffuses from these as well. The effect of a barrier upon a diffusion is shown in column (d); this may eventually be broken, but for a period there is on one side of the barrier a time-lag in the concentration of the features being diffused. Column (e) introduces a more complex set of ideas, for beginning with an existing distribution, new elements or arrangements are introduced, first appearing in the north, and then gradually spreading south as a wave of innovation, diffusing through and within the existing distribution until the new elements have replaced the old in all but a few residual cases. While the 'wave' idea was originally developed to account for the spread of consumer goods, for example car ownership in Sweden (Hagerstrand 1952), it can be adapted for understanding the complexities of settlement evolution. Indeed, settlement, developing through very long periods of time and vast areas of space, involves all types of diffusion, including the final case, (f), spread along lines of communication such as rivers or roads. Column (e) is a particularly sophisticated case because it involves contraction as well as diffusion. The two large-scale examples shown in Figure 3.5 may be explained in terms of both core–periphery models and many types of diffusion process. Indeed the USA, because it shows a documented chronology of territorial acquisition on a large scale, gives an impression of a vast, sustained wave of land-taking. Nevertheless, in detail the movement of this frontier was infinitely complex, creeping forwards, sometimes slowly, sometimes rapidly, reflecting the varied processes of the model in Figure 3.6.

This view of the settlement of the USA permits a further layering of complexity to be suggested (Figure 3.5). In North America the colonists were moving into a 'new' continental land mass which was, relatively speaking, only lightly settled. In Europe, the growth of the states framed in Figure 3.5 had roots within a migration phase which followed the collapse of the Roman Empire (Figure 3.6) and led to an inward movement of peripheral barbarian peoples. This was, in the succeeding millennium, followed by development phases which saw not only the peaceful activities of colonists but the passage of armies in numerous wars which diffused people and ideas across the continent's face. Figure 7.4 is one end-product of this activity. What is perhaps surprising is that there is any coherence at all in the distribution, suggesting that large-scale underlying factors may have been operating to give form to chaos.

Hierarchical models and dependency

Very interesting links between the realities of settlement geography and models are undoubtedly to be found in the work of Christaller (in OU 1970: 362–7; Everson and Fitzgerald 1969: 101–11). Using south Germany as a case study, he created, in the 1930s, a series of models of the spatial structure of central places ranging from hamlet to town. Figure 3.7 summarises elements of these. The roots of the models are to be found in ideas of spatial competition. Market towns need customers, and on an isotropic surface, i.e. a plain of uniform qualities, not only will regular spatial patterns of settlements evolve with hexagonal territories (Figure 2.5), but an ordered arrangement of types of settlement, hamlets, villages and towns, will emerge. Given the nature of these settlements there always tend to be more hamlets and villages than towns, and, at least in theory, more hamlets than villages. As Figure 3.7 shows all can be arranged logically within a hexagonal lattice representing varied types of territory, so that at each step in the hamlet–village–town hierarchy lesser settlements and more important settlements possess the same numerical relationship to each other, with x hamlets subservient to y villages, and y villages, together with x hamlets, subservient to z towns. Christaller defined and modelled several sets of these relationships. A small town may have six villages dependent upon it (the K = 7 lattice, i.e. six dependencies plus the one central place), a logical administrative arrangement, but where transport was crucial a market town might share the allegiance of six villages (theoretically each a half share, six halves plus the central place giving a K = 4 lattice). Where services offered by a central place were unhampered by problems of administration or transport the dominant linkages were purely those of the market functions of the central places, resulting in each village being 'shared' by three market towns, the K = 3 principle (i.e. six-thirds = 2, plus the central place). The 'K' figure is a measure of the interaction involved.

Christaller's models are much described, much discussed and probably much misunderstood; they present fascinating theoretical questions and challenges, and these have excited and stimulated many enquiries. Too little, however, has been done upon their historical implications. First, while they are normally described in the sequence K3, K4, K7, this author has a strong feeling that if the lattices were to be arranged in a

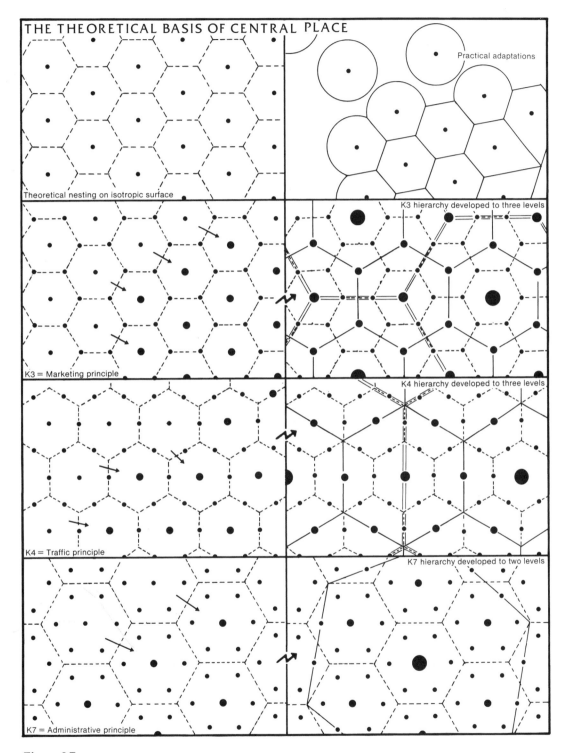

THE THEORETICAL BASIS OF CENTRAL PLACE

Practical adaptations

Theoretical nesting on isotropic surface

K3 hierarchy developed to three levels

K3 = Marketing principle

K4 hierarchy developed to three levels

K4 = Traffic principle

K7 hierarchy developed to two levels

K7 = Administrative principle

Figure 3.7

historical sequence of development, the order would be K = 7, with the administrative arrangement being the earliest, K = 4, when a transport system powerfully affects development and patterns evolution as settlement and trade develop, and K = 3, the fundamental marketing principle, being present in the developed settlement systems of the nineteenth and twentieth centuries. Second, it ought to be possible to apply this analysis within a time context, but this is in practice difficult. Work in England suggests that as a general rule, the hierarchical relationships involved often seem to have appeared before 1200, pushing the research problem to a period when evidence is either absent or particularly difficult to assemble. This conclusion no doubt also applies to south Germany − a landscape of ancient settlement − but the process of development from relatively unspecialised agricultural hamlets to a system containing numerous hierarchical levels is more detectable in other parts of the world.

The fact is that to understand these models demands experience, imagination and considerable caution, as can be illustrated with reference to Figure 6.2, which introduces the idea of change in settlement patterns through time. The base of the column represents an area or region at one point in time, call it 'T$_1$'. As time passes, some settlements survive, some are destroyed and some new ones appear; thus, the pattern seen at the top, 'T$_3$', though descended from what was present in 'T$_1$', is different. The 'time plate' labelled 'T$_2$' merely maps all of the settlements deserted at about that time − why this should happen does not for the moment matter. The important idea is that the study of settlement involves stability, contraction and destruction, expansion and addition.

Pattern characteristics: random, regular and clustered

Regular, clustered and *random* are words which have been used to describe the spatial geometry of a given settlement pattern (Figure 2.1, lower), and define conditions which are measurable by nearest neighbour analysis. Varied methods of calculation of a nearest neighbour index (NNI) have been much described (e.g. Haggett *et al.* 1977: 98–102). The end-product is a figure which ranges from 0, where the points are completely clustered together, to 1 where the points are completely random, to 2.15 where the points form a completely regular distribution. This technique must be approached with considerable caution, for several reasons; it effectively conceals internal variations within a single pattern selected for analysis. For example, the historical analysis of settlement shows that the entities present within a single distribution often incorporate not one but several different distributions, each dating from a different time period. Further, even if the end-result is close to 1, i.e. random in geometric terms, the disposition of settlements in the real landscape may not be random, but may correlate directly with, let us say, the distribution of a given type of soil. Finally, scale is crucial, so that a distribution viewed at a scale of 1:50,000 may be highly clustered, while the same distribution viewed at a scale of 1:10,000 may be random. In fact the index is best used by way of comparison, selecting identical sample areas from several regions. This does raise interesting questions, for the technique is descriptive, not explanatory, and in practice 'eyeball' methods will reveal the essential contrast between nearest neighbour indexes of 0, 1 and 2.15, while the difference between 1.1 or 1.89 and 2.05 may have little geographical significance, for at most scales there are problems of accuracy of measurement.

Nevertheless, when examining settlement patterns over large areas, for example in Figures 3.1, 3.2 and 3.3, spatial regularities and irregularities do undoubtedly appear.

Figures 6.5, 6.6 and 6.7 show the same tendencies. The lower section of Figure 2.1 is a diagrammatic representation of varied types of settlement patterns, forms and combinations. The patterns modelled are wholly theoretical, a reference framework from which to create a more precise terminology, but they serve to emphasise the varied ways in which settlements, be these single farmsteads, hamlets, villages or towns, can be arranged in space. More usefully, the insets within Figure 2.5 use the same idea in a slightly different way, and by emphasising a range of possibilities reveal the role of localised resources in generating distinctive territorial patterns. On a plain of uniform agricultural potential (a), in order to make maximum use of the arable resource, settlements will tend to adopt a regular distribution, with their territories approximating to hexagons. The appearance of a localised resource on this otherwise uniform plain (b) – for example, a peat bog or area of grazing land – will distort the territories, while the presence of only a finger of good arable – for example a valley in an upland area – creates a marked distortion in both the overall pattern and the arrangement of the territories. Alternating bands of good land, poor land and wet land help to generate strip-like territories (c), which may become even more exaggerated when settlement clings to a road or levee (d), while in extreme circumstances all settlement clusters tightly at one location, near an area of good arable (e) or a hilltop for defence (f). These are rather elemental cases, but they raise important questions: for example, in case 2.5c, can a series of strip territories set across the grain of the land to gain access to diverse resources arise spontaneously, or must a degree of careful planning be present?

Dependency: estates and boundaries

Once settlements are no longer isolated subsistence units, with farmers living on what they produce, the idea of dependency becomes critical in understanding them. Before the rise of a fully developed market economy surpluses were paid as rent to rulers or landlords. This could either mean that the surplus grain and stock were eaten by a peripatetic aristocratic household – for in England even Norman kings and their households ate their way around royal lands – or the goods were carted, or walked, to a central point for consumption. As market systems evolved, and as kings encouraged and guaranteed the circulation of stamped and authenticated bullion – money, a standard medium of exchange – stock or other produce could be sold in local markets to raise funds to pay rent in cash. The rise of a cash economy and the evolution of intra- and inter-continental marketing systems have eventually created vastly more complex bonds, so that today throughout the developed world all farms are now dependent upon the outflow of saleable produce and the inflow of ideas, machines and fertilisers. Nevertheless, throughout time, some settlements have always tended to become more important than their neighbours, and this is the ultimate root of the concept of central place which underlies the ideas developed by Christaller and Garner's premises.

Dependency and landownership run in parallel, for a powerful landowner has control over the people and the resources of a large tract of land, i.e. an estate. It is still possible in England, indeed throughout northern Europe, to see in the landscape the contrasts between 'closed' estate villages, under the control of a single landowner, and 'open' villages, usually dominated by freehold farmers. To this day, historical circumstances will often mean that estate villages remain poor in services, while open villages, where enterprise could thrive and population increase, are often larger, more populous and

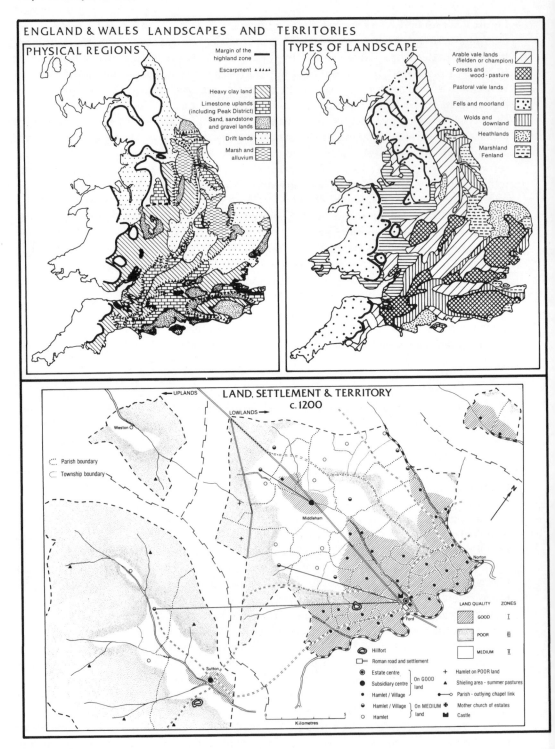

Figure 3.8

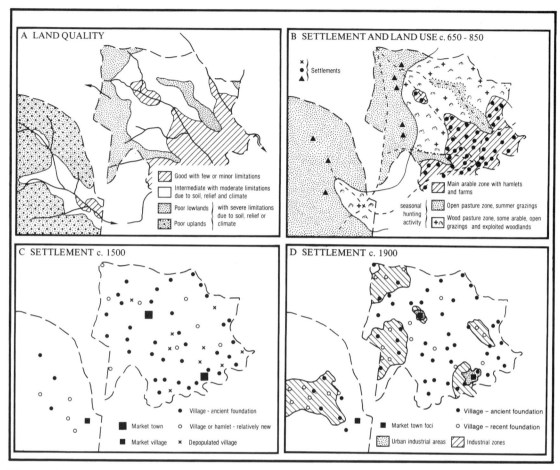

Figure 3.9 Land settlement and territory: a model

with traces of the former presence of numerous small shopkeepers and tradesmen. Estate villages may still be dominated by the 'great house', the present or former residence of the landowner. The importance of landownership as a mechanism controlling settlement is illustrated in Figures 3.8 and 3.9 which draw together into several diagrams a lot of ideas. The territory shown in these two figures represents an estate, i.e. land under the control of a single owner, and while this model is based upon English evidence it is of wider application throughout Europe. The basic agricultural quality of the land is mapped in Figure 3.9a. Of course, the two blocks shown here as contiguous may in practice often be separated by several miles. Each portion of the estate includes varied types of land, but one section is predominantly lowland in character, while the other is upland. This was a deliberate arrangement because each section then possessed a varied resource base, necessary for the maintenance of a seigniorial household. Figures 3.8 and 3.9b suggest that the territory was exploited in prehistoric and Roman times, and by the Anglo-Saxon period, say between about 650 and 850, the estate was already in existence, with well-defined regional patterns of exploitation closely adjusted to land quality. The large map 3.8 (lower) provides a more detailed glimpse of the fabric of settlement, such as might have developed by about 1200. Villages and hamlets were then concentrated on the better-quality land, with smaller clusters appearing in the wood-pasture and open-pasture zones of the estate, where grazing lands and woodlands played a more important part in the economy. By 1200 the land was already divided into territories: townships, the basic units of land attached to each particular settlement, and parishes, where the support of a church was involved, some parishes being a single township unit while others are comprised of several townships. The details of this arrangement may be compared with the stylised versions given in Figure 2.5, with the smaller parishes and townships on the good lands, strip parishes and townships responding to the banding of land quality, with very large territories in areas of poorer land. On maps the presence of such place-names as Norton, Sutton, Weston, and Middleham, reflect location within the estate, north, south, west or middle. The division of the estate into two sections, one dominated by good-quality land, the other by poor-quality land, once gave the owner access to grain-producing and pasture areas, but today some of the fine lowland acres may form parkland around a great house while the uplands serve as grouse moor.

Figure 3.9c is a skeletal plot of the types of nucleus present in about 1500, by which stage it will be noted that a number of settlements have been depopulated. There are in fact something of the order of 2,500 deserted villages in England, ranging in date from the Anglo-Saxon period to the nineteenth century. The processes of settlement depopulation were most intense in the fifteenth century, a time when changes in farming encouraged landlords to eject their tenants to create great sheep runs. Figure 3.9c shows that these tend to concentrate in good soil regions, zones normally dominated by mixed arable farming based upon open, communal townfields, where depopulated villages and hamlets represent a striking anomaly. The final map (Figure 3.9d) is a reminder that the underlying mineral riches of some estates were such that their owners eventually encouraged the development of areas of industrial settlement. Of course, by this stage it would be incorrect to portray many estates as wholly consolidated units, for, through time, portions were severed and added. Nevertheless, to understand much European settlement the role of great estates is crucial.

Types of countryside

The Elizabethan topographer William Harrison had a shrewd eye for countryside:

> it is so, that our soil being divided into champaine ground and woodland, the houses of the first lie uniformlie builded in every town togither, with streets and lanes; whereas in the woodland counties (except here and there in great market towns) they stand scattered abroad, each one dwelling in the midst of their owne occupieng.
>
> (cited by Homans 1941: 21)

In this account he links together several types of countryside, champion countryside, then dominated by open communal townfields and mixed farming, and woodland areas, dominated by hedged small fields. The former landscapes were characterised by nucleated villages and the latter by dispersed farmsteads, each set amid its own fields. The distribution of these types of countryside in England and Wales is shown in Figure 3.8, alongside a simplified version of a map of physical regions; both maps should be compared with Figure 3.2. To champion and woodland landscapes have been added a third, open pasture, the upland regions of western and northern Britain. Landscape types, products of land, culture and time, reflect deep-seated physical conditions. Together with superimposed patterns of territoriality, townships, parishes and estates, they create frameworks within which settlement has evolved (Figure 3.2).

Rural settlements in the Netherlands

The set of four maps represented by Figures 3.10 and 3.11 provide a picture of a national distribution for which a concise analysis is feasible because data exist to explain both its historical development and contemporary characteristics in remarkably precise terms. Figure 3.10 categorises the visible elements of the rural landscape, settlements and fields, on the basis of relative age. The country is largely dominated by old cultural landscapes, medieval in date, i.e. whose essential characteristics were generated before 1500, although, and this must be stressed, these have been altered and adapted by later and contemporary usage. Two phases of newer landscapes have been defined: first, those from the period between 1500 and 1850, comprising drained lake areas, lands reclaimed from the sea and inland areas of former peat cutting; the second category comprises those areas, either former heathlands or endyked polders, reclaimed between 1850 and 1980. This coarse division omits much detail, but the broad pattern of development is clear. Basically the Netherlands can be divided into two parts, the high Netherlands to the east and south, a region of sandy soils and heaths, and the low Netherlands to the west, never more than 1 metre above mean sea level and the lowest point being 6.7 metres below this datum, a region with clay and peat soils, and once a zone of shifting relationships, where land, river and sea graded gradually into each other. The great achievement of the period since 1500 has been to protect these lowlands by sea-dykes and reclaim large areas of the Rhine–Maas–Scheldt delta and even the Ijssel Meer, the former Zuyder Zee. The name-change was occasioned when this inland water body was converted into a lake by the construction of a broad dam between 1927 and 1932. Of course, there were medieval reclamations, between 1200 and 1500, particularly in the delta, and these are visible on the map as the extension of 'medieval' landscapes into that region; while limited in area their presence is, nevertheless, closely reflected in the details of settlement.

The map of rural settlement types is much more difficult to explain concisely; three core zones dominated by a mixture of villages (more than eight farmsteads), hamlets (two to eight farmsteads) and single farmsteads appear, in the north, in the centre and in the south of the country. This generalisation conceals a vastly complex landscape (Steeg 1985; Smaal *et al.* 1979; Riedé 1987), the product of long centuries of adjustment and maturation, where ancient manorial farmsteads – manor houses – formed the focuses for the development of large villages of tenants, with peripheral hamlets developing on former pastures and heaths during the population peak of the thirteenth century. These three zones are really an extension of a great central European region dominated by irregular agglomerations of village and hamlet size (Figure 7.4), core areas of deeply rooted medieval settlement. However, an attempt to compare this detailed map with the generalised continental version reveals the practical problems of classification and boundary definition, for different scholars place varied interpretations on the data they have available, so that the two maps inevitably differ in detail. Neither is 'wrong', but

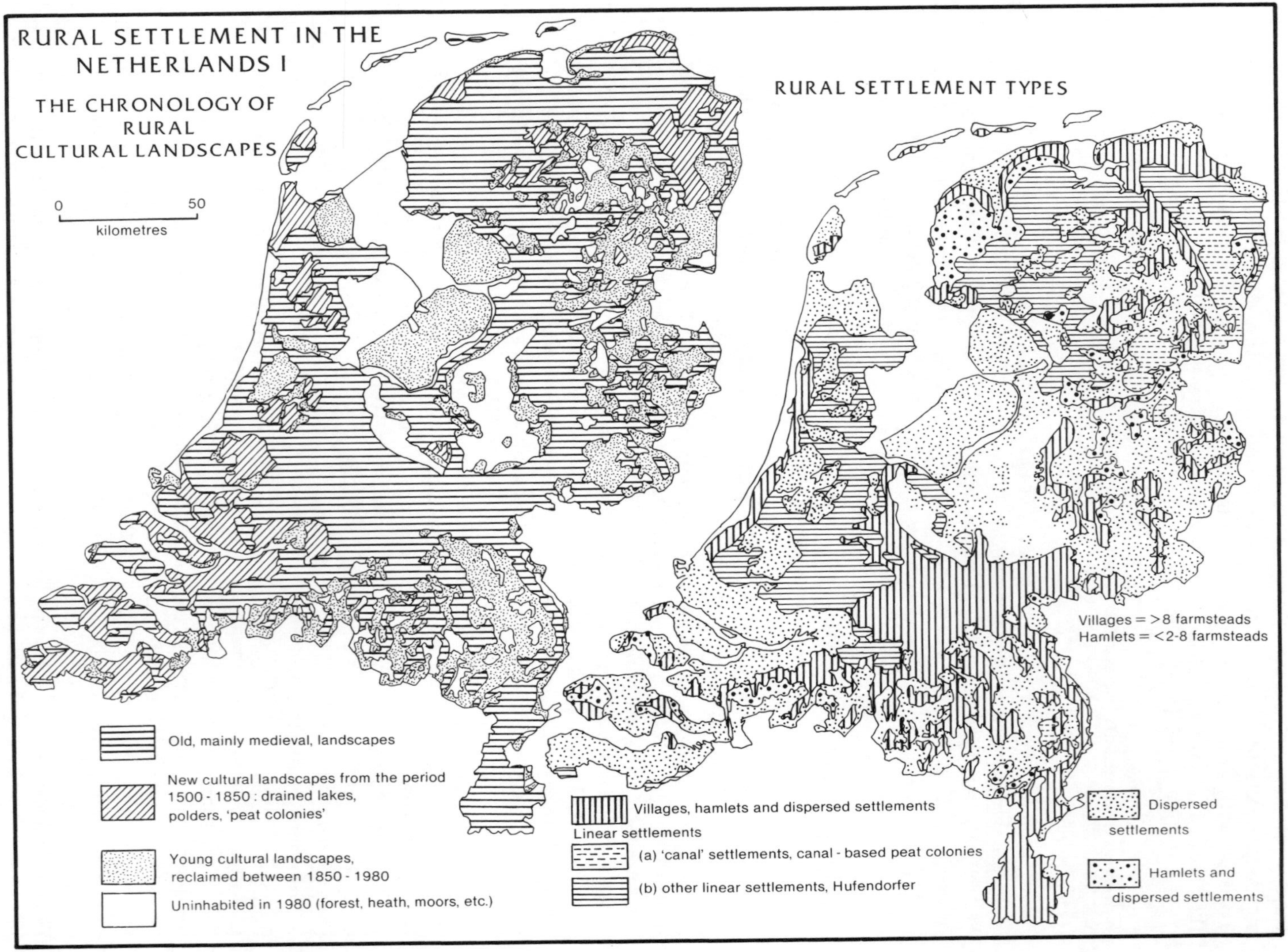

RURAL SETTLEMENT IN THE
NETHERLANDS I

THE CHRONOLOGY OF
RURAL
CULTURAL LANDSCAPES

RURAL SETTLEMENT TYPES

0   50
kilometres

Old, mainly medieval, landscapes

New cultural landscapes from the period
1500 - 1850 : drained lakes,
polders, 'peat colonies'

Young cultural landscapes,
reclaimed between 1850 - 1980

Uninhabited in 1980 (forest, heath, moors, etc.)

Villages, hamlets and dispersed settlements

Linear settlements

(a) 'canal' settlements, canal - based peat colonies

(b) other linear settlements, Hufendorfer

Dispersed
settlements

Hamlets and
dispersed settlements

Villages = >8 farmsteads
Hamlets = <2-8 farmsteads

Figure 3.10

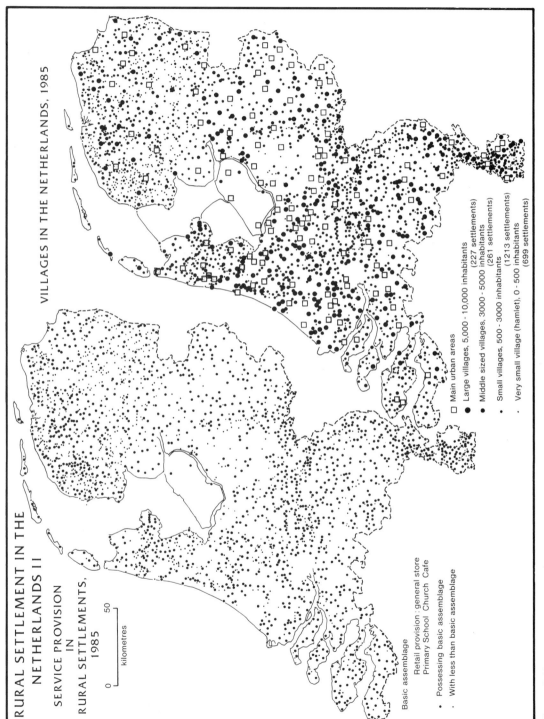

RURAL SETTLEMENT IN THE
NETHERLANDS II

SERVICE PROVISION
IN
RURAL SETTLEMENTS,
1985

VILLAGES IN THE NETHERLANDS, 1985

0   50
|_____|
kilometres

Basic assemblage
    Retail provision : general store
    Primary School Church Cafe

• Possessing basic assemblage
· With less than basic assemblage

☐ Main urban areas
● Large villages, 5,000 - 10,000 inhabitants
                                    (227 settlements)
● Middle sized villages, 3000 - 5000 inhabitants
                                    (261 settlements)
• Small villages, 500 - 3000 inhabitants
                                    (1213 settlements)
· Very small village (hamlet), 0 - 500 inhabitants
                                    (699 settlements)

Figure 3.11

both are subject to limitations. It is, however, valid to see the three core areas as zones from which two frontiers were attacked, one looking seawards, to the clays and peats of the higher portions of the low Netherlands, where settlements of the earlier reclamations are strung along the dykes and canals of the drainage systems, and the areas of systematic peat cutting. They normally assume linear forms, e.g. Kamerick (Figure 6.5). In contrast, both the sea edge and delta reclamations and the heathlands of the high Netherlands tend towards hamlets and dispersed settlement. A second frontier looked inwards, to the heathland tracts of the interior.

To counter-balance these generalisations, Figure 3.11 shows every village in the Netherlands, including a reminder that this densely populated country has many towns – indeed in the Netherlands the rural/urban population is divided 5:95 per cent. Nevertheless, the map reveals well-marked spatial variations in the patterns of villages and hamlets, here graded on the basis of population. Whatever their origin, they are now integrated within a core–periphery framework, with a great central wedge of large villages in the western Netherlands, in the eastern portion of the Randstad, where the overall population per square kilometre reaches 909 persons, although in portions of the district of south Holland, including Rotterdam itself, this reaches 1,064 per square kilometre. In spite of the concentration of villages in Limburg, the southern extension of the country, the overall density is only 490, but this contrasts markedly with the eastern and northern Netherlands, respectively 278 and 184 persons per square kilometre, but of course the density of towns is the controlling factor. More particularly, using Limburg as a base for comparison, where approximately 58 per cent of the inhabitants live in places with below 10,000 people, in the more rural northern and eastern provinces this figure ranges between 35 and 60 per cent, generally being around 40 per cent; in contrast, in the western provinces the figure is generally below 25 per cent, rising again to over 50 per cent in the delta.

These measures of rurality are reflected in local service provision, also seen in Figure 3.11; the basic package of general store or small supermarket (selling bread and meat), a primary school, church and café (an interesting selection, for a garage might be seen as equally important) is present in 58.9 per cent of the villages (in all 1,411 cases), in part present in 33.5 per cent (802), and absent in 7.6 per cent (182 cases). However, it should be appreciated that the wealth of the Netherlands, the ease of communications on a good well-ordered road system (in spite of the traffic jams in and around the Randstad), ensures a remarkably even spread of services 'where the people are', so that internal contrasts are subtle and blurred. This is a real pointer to the differences between the rural landscapes of the west and those of less developed parts of the world.

Conclusion

The detail appearing in the four national maps of settlement is such that each could warrant book-length analysis, for none can be satisfactorily described, still less explained, within the compass of one chapter. However, in the remaining chapters, a sharp change of scale will first take discussion into the family space of the individual house, where fundamental decisions are so often made, and then through the communal and private spaces of settlement forms, to their amalgamation into patterns which can be examined and sometimes explained at the workaday scale of 1:50,000.

CHAPTER 4

House and farmstead

Buildings represent much in human society: primarily they are homes, where individuals are born, live and die, but they are also places of work, places of recreation and places of storage. When analysing the actual physical characteristics of individual settlements – and indeed individual buildings – it is helpful to envisage the presence of three types of space: first, there are *public spaces*, areas where everyone has a right to go, both the individuals inhabiting a place and their neighbours, together with travellers from other areas. Such spaces are characteristically seen in highways, streets, lanes and footpaths. In most European settlements these are closely defined, but in other regions of the world defined routeways merge with the *communal spaces*, i.e. the open land where the settlement is placed, where stock and people move freely providing they do not intrude into private dwellings or yard areas. Throughout Europe these spaces survive as commons or village greens, set within, near or around a settlement, although in many upland areas open lands close to the settlement merge with larger zones of unenclosed rough grazing land. However, clear-cut as this division between public and communal might seem, Pina-Cabral notes (in Layton 1989: 61) that in the Alto Minho area of north-west Portugal the *caminhos*, old paths giving access to the settlements, were watched and controlled by the local community, effectively preventing the entrance of undesirable strangers, something which cannot be done with new paved motor roads. This leads to an interesting idea that the public spaces of roads represent for many settlements an unwelcome intrusion, and suggests that in many areas deep and subtle forces may have been at work differentiating between those settlements which were subject to easy open access, and those where access was more difficult. The cause may be no more than a few kilometres' difference in location, but the effects, open contacts in one place and closure and introversion in the other, can be profound.

Finally, there are the dwellings and areas of *private* land closely associated with the public and communal spaces, either enclosed or open, house plots, gardens or field areas. The three types, public land, communal land and private land, define categories of structural elements which, in distinctive combinations, construct settlement forms. They are present in all but the most simple settlements and together frame settings for the dwellings and other buildings. Figure 4.1a contains diagrammatic definitions. However,

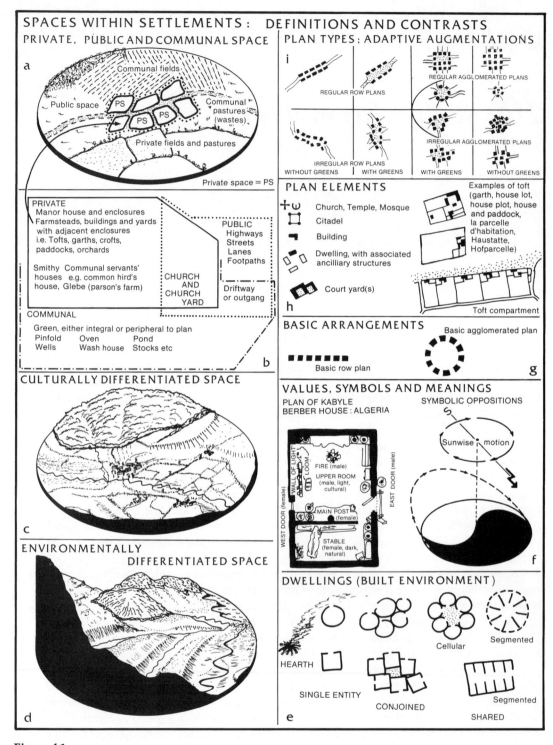

Figure 4.1

one anecdote will illustrate a most crucial point about settlement space, and emphasise that within settlement space has different qualities from space without. One snowy night this author was driving from his home, eastwards, across a ridge of high land; the snow was deep, several feet in fact, but a snow-plough had been through, making the roads into narrow, deeply cut hollow ways. There were few villages along the route, so that the relief in eventually reaching a village, with less deep snow – itself an interesting and revealing observation concerning site selection – but with the lights, warmth and humanity represented by the dwellings, was particularly powerful. The poet Andrew Young caught this deep feeling when he wrote:

> No, not tonight
> Not by this fading light,
> Not by those high fells where the forces
> Fall from the mist like the white tails of horses.
> From that dark slack
> Where peat hags gape too black
> I turn to where the lighted farm
> Holds out through the door a golden arm.
>
> (Andrew Young, in Lowbury and Young 1985: 43)

These simple words capture a deeper ancient fear of nameless things, walkers beyond the rim of the firelight. Plunging waterfalls become nightmare creatures of other worlds and times beyond this time. In the light they offer no threat, but tired, in darkness and mist, wet and cold, we still struggle against them. In former times such fears have been real and sharp for most human beings, indeed in many countries they still are, but hearth, home and firelight afford some protection.

Of course, a prosaic classification of settlement space is quite feasible, and this appears as Figure 4.1b in a diagram devised for a classic European village or hamlet, so that features such as a church, glebe or parson's farmstead, manor house and pinfold for straying cattle appear, but the underlying framework is more general, i.e. private space, communal space and public space (Roberts 1987a: 2.2). These distinctions are part of an invisible mosaic imposed upon the land beneath, interlocking in complex ways with the physical aspects of any settlement, the buildings, the enclosed areas and the open areas. Chapter 5 will deal with the varied and distinctive ways in which these, the visible and the invisible, are arranged within settlement forms. Figures 4.1c and d are a further reminder that we experience space in two ways; on the one hand there is environmentally differentiated space, and on the other is culturally differentiated space.

Buildings are the most important visual cultural manifestation of settlement and the sequence of illustrations, from Figure 4.1e to Figure 4.1i, provides a linkage between the material presented in this chapter and that in Chapter 5 on settlement forms. The hearth represents the core of the human home, where food can be cooked and smoked, and other materials charred, softened, hardened or transmuted. The control of fire provided human communities with a powerful tool. The sheltering structures placed around the hearth may vary greatly, from the single-entity unit – the simplest round or rectangular hut – to the most elaborate architect-designed detached dwelling of many rooms and storeys; from conjoined structures, where individual dwellings jostle and touch in apparent disorder, to intricate geometrically arranged cellular or segmented arrangements within a basically round framework or formal sub-divisions within a large communal rectangular structure. There is a clear tendency for technological and economic advance to lead to the elaboration of the individual dwelling, particularly those structures of durable materials designed to last for several lifetimes or even centuries. Examples of all the types depicted in Figure 4.1e can be found in both very early or near contemporary sources. For example, the segmented round house is still current in parts of West Africa yet was found in prehistoric Scotland, while the large rectangular entity, shared, and sub-divided into units of family space, is still found in the Amazon Basin and in Borneo. It is indeed, with many transformations, the basic structure of many recently constructed dwelling units, such as the Byker Wall near Newcastle, a linear block of close-packed flats, constructed to replace former – equally close-packed – nineteenth-century terraced dwellings with small houses in eaves orientation along the street.

The internal arrangements of all dwellings reflect culture and custom, technology and wealth, beliefs and attitudes. No single diagram can be an adequate summary of this diversity, but two essential points can be made. The case of the Kabyle Berber house, in Algeria, Figure 4.1f (Oliver 1987: 161–3), reveals that while convenience may determine the essential disposition of space usage, a set of cultural evaluations is also imposed. The basic rectangle is divided about a third of the way along its length to make a dark, flagstoned stable at a lower level and a higher, better-lit, living space, its floor polished with black clay and cow dung. A main ridge-beam spanning both spaces is supported at the divide by a pillar. In Berber symbolism the protecting strong beam is male, the pillar is female, while the dark end of the dwelling, where the water vessels and animals are placed, is associated with sexual relationships, birth, sleep and death – with what is 'animal'. In contrast the upper end is linked with all that is noble, the honour of the household head and the protection of his wife's virtue. This division also extends to a division of the upper end of the house, so that there is a 'wall of light' or east side of an internal wall, where the loom is placed, where the woman sits after marriage and where the umbilical cord of her child is buried, opposed by a 'wall of darkness', where storage and sleeping take place. In essence the house echoes 'natural/cultural' and male/female dichotomies, divisions which permeate many aspects of culture and life (Figure 1.5, lower diagram).

This is but a single case from many which could have been chosen. It is a reminder to 'detribalised' western readers that their own cultural beliefs and attitudes have not been, and still are not, universal. Here the word 'detribalised' is used not in any pejorative sense, but to express what has happened in a culture where family bonds are loosened, where marriages fail almost as often as they last, and where patterns of belief are diverse and inchoate, in short societies in which social coherence has given way to logical coherence (Gellner 1988: 39–69). These are not judgemental remarks, they are made to emphasise and express differences which exist. Nevertheless, the diagram in Figure 4.1f contains a reminder that many human cultures have seen the world in terms of a dichotomy, the parts of which must be brought together to achieve unity, wholeness, and that these views often build around an awareness of cardinal orientations and (in the northern hemisphere) the clockwise course of the sun through the sky in its daily journey (Dodgshon 1987: 49–60). These points find physical expression in the cardinal orientation of both individual dwellings and even whole settlements (Chapter 5). It is this element which is perhaps the most difficult problem faced when exploring the 'foreign country' of the past, yet when dealing with inherited settlements it cannot be avoided.

In practical terms, the arrangements of private, public and communal space within settlements involve the drawing together of a finite number of functional entities, as it were the building bricks of settlement plans. These plan elements (Figure 4.1h) involve features such as church, mosque or temple, fortress or citadel, together with other individual buildings, be these dwellings or ancillary structures, whose construction, arrangement and degree of proximity define spaces within and between them, yards, courtyards, dance floors and the like. Between wholly public space and wholly private space there is what Rapoport (1969: 80) terms a *threshold* (Figure 4.3, upper right), a point or line, sometimes physically visible, sometimes not, beyond which strangers should not pass without an invitation. The English garden fence is such a barrier, yet the garden space is open for access by the postman, and for newspaper or other delivery. Nevertheless, to be in the garden without an obvious or legitimate reason invites a question and if no sound reason is forthcoming, ejection is lawful. Where there is no fence, as when bare ground, a grass area, or even maintained lawns lie around the dwellings, then the front door itself, as in Mediterranean and tropical regions, may become the essential threshold. In lightly populated Sweden, provided that no damage is done or structures erected, 'every man's law' ensures wide public access in rural areas and makes fieldwork an easy and relaxed process.

Basically there are two fundamental arrangements of dwellings in settlements; they can either be strung out in a line in a *linear* plan, a row, or arranged in a cluster, an *agglomerated* plan. The two cases shown (Figure 4.1g) are, for the purposes of clarity, particularly regular and exaggerated, but as the final diagram (Figure 4.1i) shows, by introducing an assessment of the degree of regularity present or absent, and noting the presence or absence of an interior open space or green, it is possible to create a rational classification of settlement plan types – the subject matter of Chapter 5 (Figure 5.1).

Building styles

Throughout time, the styles of traditional buildings have reflected the qualities of the local environment, from the winter snow-built igloo of some Inuit peoples to the leaf-thatched dwelling of the tropical forests, from the traditional log and turf cabins of the forest regions of North America – based, in fact, upon northern European building traditions – to the mud-brick of the Middle East and Africa. It is too easy to see these as 'of the past', yet they were often superbly adapted to the environment using easily available local building materials, and served well the social and economic needs of the communities which constructed them. Even in the face of the diffusion of European styles and materials, traditional styles often prove surprisingly resilient and, when used with understanding and subtlety, can be retained in new buildings, to give a welcome cultural depth to modern landscapes. The topic is a large one and of relevance because the houses in which people live are an integral part of their lives, and can be factors expanding or limiting the human spirit. Several categories of building traditions are recognised by architects (Rapoport 1969: 1–8) and these can be used to structure discussion. *Primitive* is a term applied to the buildings of traditional societies, where the technologies involved are the most simple and where most members of the community are capable of constructing their houses, while *vernacular* is used of the styles of more developed societies and two types may be identified: *pre-industrial vernacular* styles, where at least some of the work is undertaken by specialist craftsmen, normally a carpenter, to create the basic frame, but also a thatcher, slater or tiler to create the roof. The owner and his family may still undertake the gathering of the materials and assist with some of the more unskilled parts of the construction work. This category grades imperceptibly into *industrial vernacular* styles, where the building is undertaken by an almost full-time specialist, who still uses local materials, however, and adheres to time-honoured local stylistic traditions, so that the regional personality is expressed in the buildings. Finally, *industrial* styles are the result of the standardisation of building practice and the transportation of building materials over large distances. They are, in essence, 'the town in the countryside', for they follow characteristically urban styles. At this level design and construction are wholly the task of specialists, architects, builders, plasterers, plumbers and carpenters.

Primitive styles: the case of Africa

Twentieth-century urban societies tend to view the settlements of less technologically advanced peoples as 'primitive'; this is at best short-sighted. To understand architecture of this type places great demands upon us. We must attempt to think thoughts and feel feelings wholly alien to the materialistic lifestyles of the late twentieth century – an almost impossible task. Before looking at some examples, two fundamental points must be made. First, in the minds of traditional societies symbolism is a powerful component. In western societies this is retained, and is most clearly seen in advertising, where appeals to status, sexuality and pure greed are concealed behind humour, advice and promises of pleasure, but the symbols used are constantly created from everyday objects such as cars and even electrical goods. However, in other societies, of both the past and the present, the language of symbols was one way of communicating. The churches of western Europe are full of carvings, paintings and glasswork bearing designs which carry messages whose meanings are now often wholly lost to the modern observer (Dirsztay 1978: 136–52). There is substantial evidence to suggest that many of the elaborate but nominally 'primitive' settlements enshrined a holistic view of humankind and nature, with house constructions, house plans, dwelling unit layouts and indeed entire settlement plans reflecting complex symbols, binding human society to cosmos and myth. As will be shown in a later chapter, such complex psychological structures are probably present in many European villages which at first sight seem stolidly functional.

Second, in this expression of relationships two elements are inseparable, the architecture of the buildings and the architecture of the territory; that is, 'primitive' architecture is not merely functional, it is a model of political, social and economic life. These traditional societies possess accounts, stories, myths, but they are also models – for they explain what is – which link together the group's social structure, landscape and architecture. Guidoni has summarised these bonds:

> The architecture cannot be considered apart from the territory of those populations whose chief mode of production is hunting and gathering; for stock-breeding nomads the portable house is the most important part of their utensils and the container for all of their personal goods. When, however, agriculture is the

predominant activity, the crucial factor becomes ownership of the land, and this exercises a very definite influence on the construction of buildings, whether collective or familial. The territory becomes divided into two parts; primeval nature and cultivated fields. The latter constitute the economic, juridical, and symbolic base for the development of the house, the village and what we can call an artificial, man-made world, one that is stable and capable of maintaining a particular socio-economic state. No longer restricted by the requirements of a nomadic life, dwelling and community houses can develop in forms that become more complex and more durable the more the two determining factors – ownership of land and social stratification – are emphasised.

(Guidoni 1975: 12)

No discussion of the African continent can ignore two fundamental points: in MacMaster's words 'few rural areas of Africa are occupied exclusively by one traditional society or devoted to a single economy. Still less are they now circumscribed units of self-sufficiency, if they ever were' (MacMaster 1975, 305). Following his argument there is a quite fundamental distinction between elements of the settlement geography which are autochthonous or indigenous, i.e. those rooted in the traditional societies and economies, and those which are intrusive, introduced from outside. It is easiest to identify recent European intrusive elements, but within African societies autochthonous and intrusive elements blend in complex ways. For example, Muslim influence is felt both south of the Sahara and in East Africa, while Ethiopia reflects the most ancient Christian influences. Second, compounds, involving associations of huts, private open spaces and communal open spaces, all delimited by varied types of barrier fences, are characteristically linked with essentially single- or dual-function buildings. The degree of complexity present will reflect social as well as economic factors, so that the compound of a monogamous man will be simpler than that of a polygamous one (Figure 4.3e). A compound surveyed in Cameroon, West Africa, composed a circle, with 'hut of man, hut of a wife and several sheep, kitchen, hut of the first wife and cow, hut of wife, storage of rope and nets, hut of wife and several goats, stable, kitchen, hut of wife and a goat, hut of wife and small indoor granary, and hut of man and two cows'. Add to this one hut in construction, a 'hut of headman with storage' set centrally, a rice granary and a millet granary, and the resulting settlement, often of the order of 20 metres across, would contain some sixteen buildings, most of them round but three square, housing about nine adults. Hunter's description of a compound in Nangodi, Ghana (see Chapter 1), emphasises how the basic cell of a few huts around an open space can be added to. In a society in which wealth took the form of cattle, the outer circle provided a defence against both human and animal predators.

The evidence of reports by anthropologists makes it quite clear that even in settlements dominated by circularity of buildings two processes are at work: on the one hand growth by accretion, with domestic sub-units being added in apparently haphazard ways, and on the other increases in size which involve planning, for there are cases in which a compound has been planned holistically, i.e. by laying out the fundamental circle. Compounds are a traditional form throughout much of Africa south of the Sahara and some of the varied house forms are summarised in Figure 4.2. It is tempting to see the buildings in the diagram, adapted from work by Gebremedhin (in Oliver 1971), as an evolutionary series, developing through time, but while there is probably some truth in

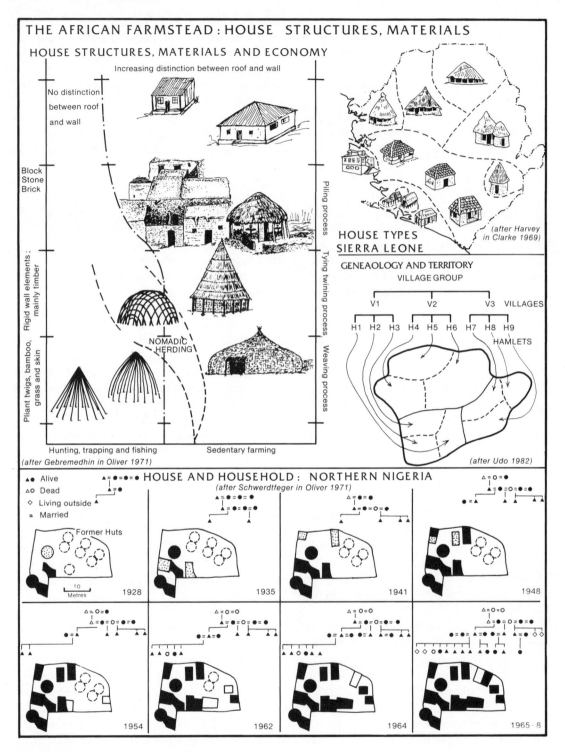

THE AFRICAN FARMSTEAD: HOUSE STRUCTURES, MATERIALS

HOUSE STRUCTURES, MATERIALS AND ECONOMY

Increasing distinction between roof and wall

No distinction between roof and wall

Block Stone Brick

Rigid wall elements; mainly timber

Pliant twigs, bamboo, grass and skin

Piling process

Tying twining process

Weaving process

NOMADIC HERDING

Hunting, trapping and fishing

Sedentary farming

(after Gebremedhin in Oliver 1971)

HOUSE TYPES
SIERRA LEONE

(after Harvey in Clarke 1969)

GENEAOLOGY AND TERRITORY

VILLAGE GROUP

V1　　　V2　　　V3　　VILLAGES

H1 H2 H3 H4 H5 H6 H7 H8 H9

HAMLETS

(after Udo 1982)

HOUSE AND HOUSEHOLD: NORTHERN NIGERIA

(after Schwerdtfeger in Oliver 1971)

▲● Alive
△○ Dead
◇ Living outside
= Married

Former Huts

1928

1935

1941

1948

10 Metres

1954

1962

1964

1965·8

Figure 4.2

this view, reality is infinitely complex. The time involved is so great that there have been ample opportunities for intermixing local traditions, and surprisingly, while broad traditions can and do survive for many centuries, wholly local styles can emerge in very short periods of time. For example, among the nineteenth-century Zulu of the eastern portion of what we now call South Africa the circle became a dominant design theme and vast but (in retrospect) ephemeral congregations of people appeared (Oliver 1971: 96–105). One such military *kraal* comprised 1,400 huts strung around a circle some 3 miles in circumference! Another royal *kraal* was said to be 'a mile in diameter, with the interior, supposedly for the monarch's cows', which were in fact kept in smaller enclosures around the edges. The open interior was used as a parade ground (ibid.: 98–9). This is a fascinating development, and an adaptation of the basic compound idea – in essence a family homestead – to a settlement form with 'urban' implications, used as a base for raids by highly organised regiments or *impi*.

A second building tradition present in Africa involves the use of mud-brick, i.e. mud mixed with water and often a vegetable binder and sun-dried to create a durable building material. This technique has ancient roots, for it was used to construct the buildings of the earliest cities in the valleys of the Tigris and Euphrates, in Palestine, in Anatolia and in Egypt. The technique is undoubtedly associated with traditions of constructing round buildings (e.g. Figure 1.1), but when used in rectangular or sub-rectangular buildings allows the possibility of adding rooms (Figure 4.1e). Arrangements of rectangular buildings around small courtyards, as among the Ashanti of northern Ghana or Hausa of northern Nigeria, taking the form of timbered walls plastered with clay, can be found in either dispersed farmsteads or repeated again and again to give closely textured urban aggregations. The exteriors are often smoothed, plastered and highly decorated, for buildings of thatch, timber and unbaked clay require constant maintenance and frequent rebuilding. Schwerdtfeger (Oliver 1987: 58ff.) has shown, using a series of magnificent drawings and meticulous family reconstruction (Figure 4.2, lower), the way in which a single compound changed over a forty-year period in response to changes in family size and fortune. Where the building materials are mud-brick and where settlement clings to the same site, then the vast and ancient tells or mounds of the Middle East, as at Jericho, can arise. The accumulation of non-transportable wealth, particularly grain, by such communities encouraged repeated *in situ* reconstructions, generating a mound of human-made debris upon which the latest settlement is sited.

Morgan presents what is an important general conclusion, suggesting that in such 'circumstances the entire "built environment" is being replaced over a short cycle' (Morgan 1983: 69). Buildings represent 'working capital' rather than 'fixed capital'. It is often easier to replace wholly rather than preserve them, one factor generating instability in the overall settlement pattern as well as a tendency for individual forms to mutate very rapidly. Change within a compound – however it be constructed, of vegetable materials, animal dung or clay, or mud-brick – is generated by the birth, maturation and death of family members (Figure 4.2, lower), leading to the creation and destruction of structural sub-units at the behest of necessity or decay, disease or infestation, and the vicissitudes of good and bad fortune. Udo (1987: 47) emphasises the close relationships in Africa between territorial and kinship organisation; the essential elements of this symbiosis are shown diagrammatically in Figure 4.2. A village group occupies a large but seldom compact area, and its unity depends upon a common name, usually that of a mythical ancestor, a common territory and common customs and taboos. Among the Ibo of Nigeria the village group was traditionally the highest political and social unit, and in pre-colonial days had only limited social links with other similar groups, although periodic markets provided economic links. From the founder of the whole group were descended the founders of individual villages, each possessing a defined territory, and each village in turn is made up of smaller units, hamlets, which may either be a separate entity or merely a section of a large compact village. Within this framework, individual settlements could be either ephemeral or semi-permanent, but the village group territory tended to have a longer duration in time and be a quasi-permanent component of the system.

Although the environment is different, this discussion of Africa provides some general lessons for Europe, a glimpse of the forces affecting the prehistoric and medieval antecedents of present settlement. Above all else it is probably landownership and a farming system using rotations to sustain soil fertility which are the two most powerful forces generating stability in settlement forms and patterns. This tendency was confirmed and reinforced by the gradual adoption of long-lasting building materials, stone and then brick.

Landownership involves closely defined proprietary rights by an individual, a community or a corporation over a defined tract of land. In contrast, unstable settlement forms and patterns tend to be present in communities, tribes or clans whose buildings are difficult to maintain, whose farming system is one where maintaining soil fertility was difficult and whose economy was based upon cattle, gold and women as movable wealth. The impact of these ideas upon understanding settlement patterns will be reconsidered in Chapter 6.

Pre-industrial vernacular: Europe, Japan and South Asia

Vernacular architecture is a style usually associated with communities of farmers bonded to the land, i.e. where landownership, involving both personal possession and family rights, has become appended to particular land areas. Demangeon summarised concisely the function of these buildings: 'rural residences exist to satisfy the everyday demands of the farmers, to give order to the foundation of rural life, and to provide the needs for various social functions' (in Wagner and Mikesell 1962). The farmstead always represents a large investment in fixed capital, and imposes a substantial burden on a nuclear family. Once constructed, it possesses inflexibility of adaptation to change and contains large amounts of movable wealth in the form of basic tools, stored crops and housed animals, and luxury goods. Thus, traditional forms evolve, adapted to culture and environmental circumstance, and once developed local styles can persist for many centuries. The subtle variations which evolve in response to local needs can be illustrated with reference to two widely separated and contrasting regions, northern Europe and Japan. Figure 4.3 contains a simple model by Rapoport of the contrasting privacy realms in a Japanese and a western – Anglo-American – house. In a normal western house of the twentieth century the barriers to intrusion from the outside are surprisingly open – generally low fences, walls penetrated by many light-giving windows – but the internal privacy is extreme, the quality of housing now being assessed in terms of provision of rooms for the individual members of the family. In Japan, in contrast, the dwelling house

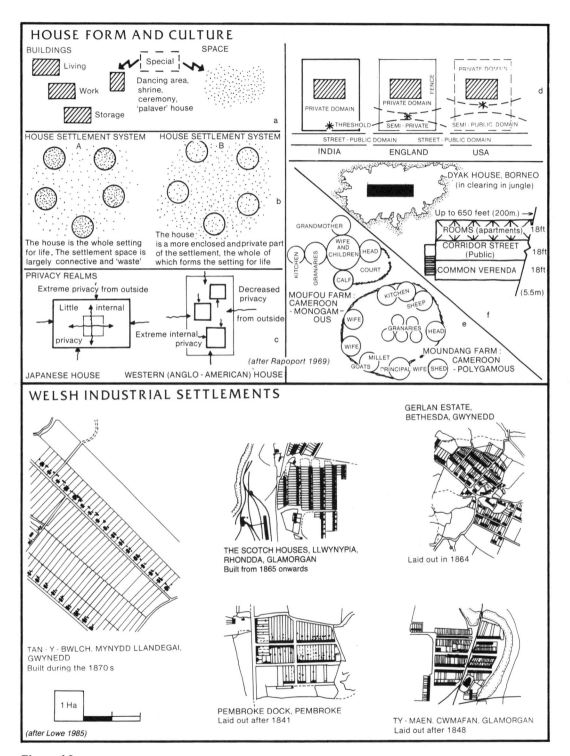

Figure 4.3

turns a blank facade to the outside world, while internal privacy is traditionally extremely limited, with walls often, quite literally, being paper thin. In former centuries, western dwellings allowed far less privacy, indeed privacy is a privilege which diffused down the social scale, from lord to commoner. Absolute privacy is largely a phenomenon of the second half of the twentieth century.

Europe

The roots of the advanced peasant houses of western Europe have been traced archaeologically (Chapelot and Fossier 1985: 72–128). The centuries after the collapse of the Roman Empire, i.e. between about 400 and 1000, saw the development of sturdy timbered dwellings using the region's rich forest resources. Stout post-framed buildings gave a rectangular shape; in humbler cases these were rather long and low, but aristocratic examples comprised 'high and horngabled' timbered great halls. Walls were boarded, often with oak, and perhaps intricately painted, while the surviving structures, invariably churches, suggest that the higher-status buildings were elaborately carved and had shingled roofs, i.e. hundreds of small tiles split from logs. It is probable that separate function structures were being created from earliest times – the separation of the precious granaries from the fire risks of the living quarters was a first step – but at the level of the normal farmer, wealthy or poor, man and beast shared the same roof, often indeed the same room. As a concession to hygiene the byre end, housing the precious cattle, lay downslope, with a drain to remove the urine. Such single-storey long-houses are discovered on sites throughout the northern European plains, from Germany to Scotland and Ireland, but differentiation into other types, essentially groupings of single-function buildings, depends upon several factors.

The accumulation of wealth was undoubtedly a prime motive, and the tradition mentioned above, of creating separate function buildings, for example hall (a general living room), kitchen, byre, stable, barn, chapel, privy and the like, arose for several reasons. Fire, as already indicated, was perhaps the most crucial, for hygiene and privacy were not highly rated (although these accounted for the separate privy or toilet), but a high-status farmstead contained accumulated wealth, grain in the bins and seed corn in the granary, smoked meats and cheeses, nuts and dried fruits, honey and herbs, hay for stock feed, leather goods and textiles, wooden implements, pottery and the products of the black-smith. In addition there would have been luxury goods, either won as loot in war or carried by local and long-distance trade, jewellery, weapons and horse trappings, fine cloths – including even silk – vessels of fine woods and precious metals, rich furs and horses, dogs and hawks. Status closely followed wealth. This involved keeping and rewarding retainers, entertaining guests and using the bounty of daughters to create advantageous marriage alliances, and engaging in war to keep young men occupied, for coercion was a depressingly common ingredient of these agrarian societies. Finally, the advantages of alliance with the church – whose successful magic was sustained through literacy and shadowy linkages to a Roman past – made keeping a priest a desirable aristocratic activity and led to the adoption of separate buildings for ritual and ceremony as well as specialist storage and function. Large complex aristocratic farmsteads formed seed nuclei from which, in later centuries, hamlets and villages grew. Nevertheless, echoes of this deeper past are still to be seen in European landscapes in the form of church-hamlets, a great house with church adjacent and nearby the dwellings of a few dependent households.

At a peasant level, where work was done by the farmer and his immediate family, there were perhaps other pressures for multi-building farmsteads. The fact that precious calves brought to birth in the warmth of the byre could be watched closely and stood more chance of survival may have been a reason for housing men and beasts under the same roof. Furthermore, the climate of the peninsulas and islands of northern Europe is sufficiently variable to necessitate inner work areas, where grain could be threshed and winnowed and where the farm could be serviced by repairs to wagons, ploughs and other tools. Thus there were incentives for those families who could to invest in substantial farmsteads, and while there were periods of famine, grinding poverty, plague, storm and frost, together with war and devastation, all was not gloom. Take, for example, the splendid paintings by Peter Brueghel the Elder, working in the middle years of the sixteenth century: their physical coarseness may not be over-exaggerated (the value of a close shave with a fine steel razor is often evident), but their material prosperity is made clear by sound clothes, swinging purses and keys, serviceable daggers, and plenty of food and drink, while in the background lie the great functional northern European farmsteads, all no doubt containing dwelling, storage and work areas, frequently showing clear evidence of two storeys and the presence of brick-built hearths and chimneys – a helpful precaution against fire. The farmsteads are timber-framed with wattled and plastered walls, thatched roofs and stout planked doors. Nevertheless, the *Massacre of the Innocents* could be, indeed was, easily converted by a few small alterations of content into the sacking of a village (Claessens and Rousseau 1987: illustrations 101 and 102), a salutary reminder of the troubled times through which Brueghel lived.

Figure 4.4 incorporates examples of farmsteads from northern Germany and Scandinavia. These are traditional forms, with ancient roots, and – although the cases used here are from museums (Uldall *et al.* 1972; Skansen 1974) – the types still appear in the landscapes as functioning farmsteads. The styles have been adapted in the last two centuries, using modern building materials, stone and brick, tile and slate, and even concrete, breeze block and composition or galvanised iron roofing. This adaptation involves constant tension between the substantial expenses of rebuilding and the disadvantages of adapting older structures. When money is cheap, and incentives are present, change is rapid, but it slows when money is expensive to borrow. Even the provision of hard surfaces around a farmstead, sufficient to bear the weight of modern vehicles and the constant action of water, organic acids, frost and friction, is a costly undertaking. Nevertheless, changing agricultural technology and the imposition of external standards, for example in dairies, are powerful incentives for change in traditional arrangements. One Cumberland farmer put the present position most forcibly: pointing to his fine great sweep of outbuildings, largely eighteenth and nineteenth century in date, he commented sourly that these were now probably worth more than his land, for they could, at that time (the mid-1980s), be converted into dwellings for wealthy commuters.

The group of farmsteads illustrated in Figure 4.4 includes a farmstead from Ostenfeld in South Schleswig, Denmark (a), a type particularly common in north-west Germany, comprising a main nave, with an open hearth at one end, common to men and animals, although the cattle were stalled in the side aisles, where drainage could be provided. Private rooms, for the owner and his family, were added at the rear in 1787, but although this particular building dates from 1685, the type has roots in the medieval and even the later prehistoric periods. It was normal in Denmark – and South Schleswig belonged

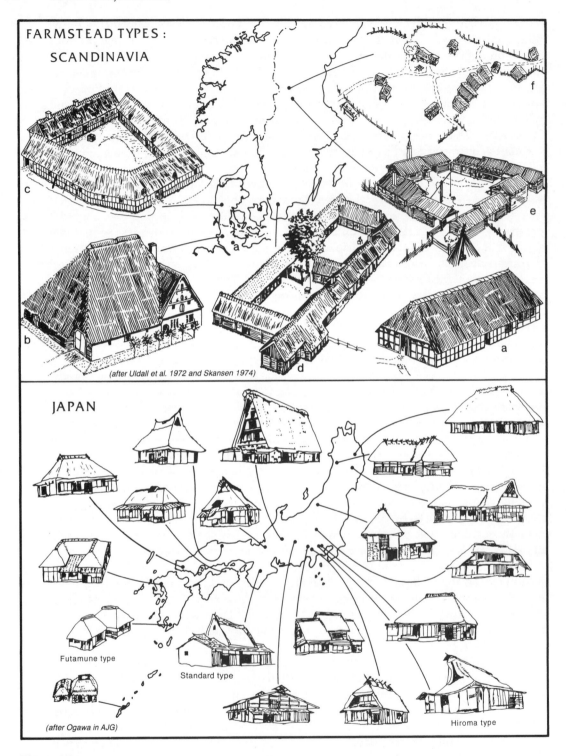

FARMSTEAD TYPES:
SCANDINAVIA

(after Uldall et al. 1972 and Skansen 1974)

JAPAN

Futamune type

Standard type

Hiroma type

(after Ogawa in AJG)

Figure 4.4

to Denmark until 1864 – for this type of house to be orientated east to west, with its main entrance on the village street, a tendency linked with a north to south street orientation and perhaps village planning procedures. It will be seen that the timber-framed and thatched building, with a capping of heavier turfs along the ridge line, had no chimney, so that the interior beams were preserved by thick layers of soot deposited from the smoke that sifted out through the thatch. The central large room had a hard clay floor, on which threshing and other farm work could occur, while hay was stored in the roof loft, where it was kept dry by the heat from the hearth below. The fire risks were horrendous, and probably account for a common tendency throughout northern Europe to have a wet ditch around the building, often sufficiently large to be a defensive moat. A variant on this same theme is seen in the farmstead from Eiderstedt (Figure 4.4b), also in South Schleswig, a farmstead type found, with variations, extending along the marshlands between the Netherlands and Denmark. The building rises some 13 metres (44 feet) to the roof ridge, and comprises a great central barn demarcated by four massive posts which form the frame of the building. Around this core are arranged separate living quarters, cowsheds, stables and the threshing floor. This type of construction seems to have evolved after about 1600, a time when rising corn prices and increasing exports necessitated larger storage spaces. A surrounding wet ditch is characteristic of the whole zone, both serving as a drain and offering some protection against fire.

These large farmsteads, often using massive and long timbers, draw together under one roof the activities of a whole steading. That from Lundager (Figure 4.4c) is of a type found with local variations throughout Denmark, an arrangement of buildings around a courtyard, in this case not completely rectangular because of the need to adjust one wing to the line of a pre-existing village road (see Figure 5.5, Nordby). All the buildings are framed in timber, the spaces being infilled by wattle and daub or sun-dried bricks of clay. A covered gateway leads into the cobbled yard, where the well and the dung-heap were sited, an unhygienic but once very common arrangement in rural areas. Barns, one for each type of grain, stables for horses, stalls for cattle and other workshops close in the yard. The steading contains structures of many periods, ranging between the later seventeenth century and the nineteenth, and such heterogeneity, linked to changes in use, will often preserve, as a barn or stable, an ancient dwelling house.

A journey northwards through Scandinavia reveals variations in construction and subtle changes in layout. The great double farmstead from Goinge, Scania, Sweden (Figure 4.4d), contains structures dating mainly from the eighteenth century, although some elements are earlier, and the walls, in this southern part of the country, are built of heavy horizontal oak boards, with the roofs being thatched. Such buildings are often difficult to date, for any timber structure is a battleground between constant decay and constant repair, and replacements on the same foundations and the use of older materials in a newer structure are but two of the problems. The steading from Mora, north of Lake Siljan in central Sweden (Figure 4.4e), dates from the sixteenth century, and here the principal buildings are all of 'log cabin' construction, i.e. pine logs with notched interlocking corners. In fact there are two dwellings, one for the farmer and one for aged parents, a not uncommon arrangement. The final example, a shieling from Alvardalen – a mountain farmstead from Dalarna, north-central Sweden (Figure 4.4f), was occupied during the summer months to take advantage of the more distant grazings. The scatter of buildings comprises dwellings, haybarns, a barn for dried leaves – a nutritious fodder

– cowsheds and a dairy, the largest central structure. Here 'shielingmaids' lived for between six and nine months to tend the sheep and goats.

Japan

Japan presents a complete contrast to Europe. Ogawa (AJG 1980) has attempted a synthesis of the rural house types of Japan in a paper which pays careful attention to remarkably slight and subtle variations in layout. Although usually on stone foundations, the building materials are normally organic, with mud or wooden walls and light movable wooden or paper partitions attached to frames of wooden posts, either jointed or bound together. Roofs are of thatch or tile, or more recently galvanised iron. Two key elements are always present; first, the *doma*, an earth area, with no constructed floor. Here daily activities such as cooking are carried out, at the same level as the ground outside, grain is stored and horses stabled. The *yuka*, with boarded raised floorspace, forms an inner living area, never entered in outdoor footwear. In the standard farmhouse type there are normally four rooms, the bedroom for the head of the household, a living room with a fireplace, a guestroom, and a room for ritual assembly, entertaining, weddings, funerals and Buddhist ceremonies. In the largest buildings there will be sub-divisions to make six or eight rooms. A variant (*hiroma* type, see Figure 4.4) has only three rooms, a bedroom area (often sub-divided), a living room, and a room for assembly. All are gathered under one roof. As might be expected, scholars recognise a range of sub-types. Thus, in the north-east of Japan, a region of cold climate with long winters, overcast skies and heavy snow where a protected workspace is needed, the *doma* is larger. A further variant (the *futamune* type, Figure 4.4) provides separate roofs for the *doma* and the *yuka*, giving the appearance of two houses with a space between. This type appears most frequently in the warmer regions of Japan, prone to typhoons, and rather more backward than the mainland. However, it has also been argued that this *futamune* type is in fact 'earlier', so that its singularly patchy distribution probably represents residues of an older, once more widespread, tradition.

South Asia

Figure 4.5 deals with the buildings styles of an entire sub-continent, ranging from Afghanistan to Assam and from Nepal to Sri Lanka in the south. The rich diversity of building styles is generated by variations in building materials, cultural heritage, current fashion and contrasts in caste and community, from the tents and *yurts* of the Pathan and Uzbek nomads of Afghanistan to the *padi* farmers of Bangladesh and the urban modernity of Delhi. A broad contrast exists between the dwellings of north-west India, flat-roofed and blank-walled boxes, looking inwards, characteristic of arid south-western Asia, and the thatched gables of more humid areas to the south and east. Slightly divorced from the main diagram are a lower set of illustrations showing poorer dwellings, and Spate (1957: 177) gives a vivid account of the houses of the Untouchables of the south, which often form small hamlets or *cheris*, set on the edge of, or some distance from, a main village:

A typical *cheri* may consist of two rows of huts with a narrow central 'street'; in the

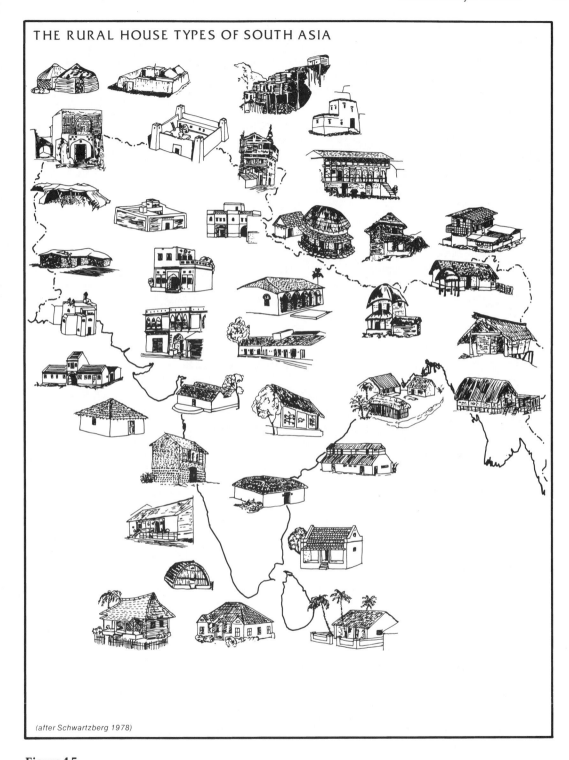

THE RURAL HOUSE TYPES OF SOUTH ASIA

(after Schwartzberg 1978)

Figure 4.5

middle this widens to make room for a tiny temple. The huts have thick mud walls, roofed with palmyra thatch, and low mud porches scrupulously swept. To enter one must bend double; the only light comes from the door and from under the eaves, and the furniture consists of a few pots and pans, a couple of wooden chests, and the essential paddy-bin, 4 to 6 feet high and 3 to 4 in diameter, raised from the ground to escape the rats, and built of hoops of mud. Poor as they are, these dwellings are yet homes, and obviously loved as such: the cleanliness, the surrounding mangoes, coconut and palmyra palms, redeem them from utter squalor. The nadir is reached in the bustees of Calcutta and the revolting camps of casual tribal labour found on the outskirts of the larger towns: shelters (they cannot be called even huts) of matting, of rags, of petrol tins beaten flat, on waste spaces open to the sun and reeking with filth.

A figure as complex as this, embracing an entire continent, defines the substance of an entire book, yet the overview, based upon the work of numerous travellers and scholars, serves to emphasise the underlying diversity, the rich cultural depths from which the present landscape rises. It invites comparison with Figure 7.5, to be discussed later (p. 164), and is a reminder of the extent to which traditional dwellings have reflected local geographical circumstances and *genres de vie*.

Industrial vernacular and industrial styles: the British Isles

Much of the architecture of the majority of ordinary buildings of the British countryside now derives from industrial roots, and these in turn are the descendants of the urbane and gracious terraced houses of mid- and late-eighteenth-century towns, although earlier examples are known – the terraces and blocks of later medieval almshouses and vicar's closes in cathedral towns such as Wells. In the growing towns land prices and the desire to live at the centre of things encouraged the construction of eaves-orientated terraced dwellings, i.e. narrow buildings, fronting the street, with their eaves and frontage all forming one line. Behind these lay the service areas, stables and servants' quarters, the mews, in the case of the wealthier structures, and small courtyards with wash-houses and privies in less affluent contexts. At first these dwellings, often of three or four storeys, were built by and for the rich, but reduced in height, vastly simplified, with no more than yard space to the rear, their debased versions are seen in the terraced ribbing associated with the growth of industrial towns, Birmingham, Sheffield, London, during the second half of the nineteenth and the early twentieth centuries. The varieties are endlessly fascinating, and are well documented, both in the original text-books of jobbing builders and in scholarly analyses (Muthesius 1982). Nevertheless, their impact upon the rural landscapes was substantial. This took place in two ways. First, there were movements for rural improvement, providing sanitised cottages and farmhouses to replace older structures (Darley 1978; Bell and Bell 1972), while, second, the urban building style, convenient and cheap, was imported into both purely rural areas and into quasi-rural industrial zones. Some examples of these, drawn from Wales, are illustrated in Figure 4.3 (lower). In rural contexts these terraces were introduced into rather remote localities, often coalfields or areas with other minerals, where there had once only been scattered farmsteads so that the building of new dwellings was necessary to accommodate the workmen needed to fill essential jobs. Throughout the Midlands, Wales and the north of England the terrace crept into the countryside (Muter 1979; Lowe 1985; Atkinson 1977).

A description of northern terraced miners' dwellings was introduced in Chapter 1 and this same style, subtly adapted to local circumstances, appeared in most of the industrial regions of England and Wales. At first, because of the cost of obtaining building materials, purely local resources were used, and structures were often close to wholly rural vernacular traditions. The improvements in transport methods brought by the Industrial Revolution, notably canals and railways, spread standardised construction methods throughout the whole country. Good grey Welsh slate for roofs replaced thatch and tile in many regions. Eventually even the window- and door-frames were manufactured in large workshops rather than by local carpenters working on the building site; standardised industrial construction had arrived. The details of individual dwellings are not for further discussion here, but the groupings of houses in the landscape are of relevance to an understanding of rural settlements. The earliest purely industrial clusters were essentially of two types. First, dwellings intruded into the fabric of a pre-existing rural cluster, infilling the gardens, orchards, paddocks and open spaces of the older nucleus, whose form could often be dimly discerned amid chaotic disorder, in which awkwardly placed buildings intermingle with meandering footpaths, there often being no sharp distinction between the public and private spaces in the settlement's interior. Second, individual industrialists or companies constructed entire settlements adjacent to mines, quarries, forges and mills and ports; Pembroke Dock (Figure 4.3, lower), begun in 1818, was extended during the 1840s and based on a grid. It originally comprised single-storey cottages. The Scotch Houses in the Rhondda valley were built from 1865 onwards to serve a colliery. Each row sits on a steep slope facing westwards across the valley, a classic example of the parallel row layout. Ty-Maen was a bleak copper miners' settlement, cut in two by a railway line, the street layout probably dating from about 1848 with house construction taking place during the 1850s, while the Gerlan estate, Bethesda, comprised about 100 plots for individual houses laid out during summer 1864. Because the buildings were approved by the Bethesda Improvement Commissioners the settlement possessed wider roads and was broken into a series of sections, each independently developed. Finally Tan-y-Bwlch, built during the 1870s, was a model settlement devised by the managers of the Penrhyn quarry. In this case, the plots were leased to workers for thirty years on the condition that they built their houses to an approved design; when the leases expired, the houses were to revert to the estate, an example of ruthless capitalist exploitation. Set on an infertile and exposed site, these upland cottages reflect a way of life, part-industrial, part-agricultural, which lay at the heart of the genesis of the Industrial Revolution.

Rural vernacular building styles: the British Isles

Figure 4.6 integrates into one drawing some examples of the regional contrasts in vernacular architectural styles found in Britain. Brunskill (1985) has pragmatically identified a series of regions in which distinctive local styles prevail. This division of the country into regions, like any type of classification, is to be seen primarily as more or less useful rather than as straightforwardly true or false, and the boundaries are only

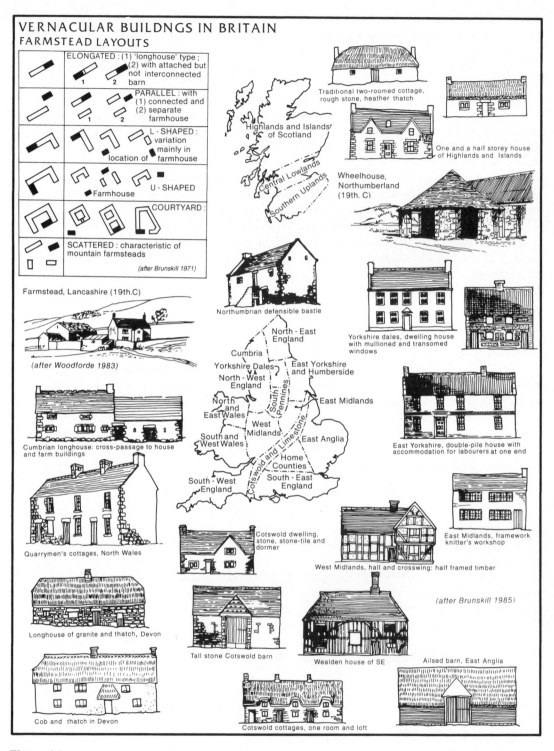

VERNACULAR BUILDNGS IN BRITAIN
FARMSTEAD LAYOUTS

ELONGATED : (1) 'longhouse' type ;
(2) with attached but not interconnected barn

PARALLEL : with (1) connected and (2) separate farmhouse

L - SHAPED : variation mainly in location of ■ farmhouse

■ Farmhouse   U - SHAPED

COURTYARD :

SCATTERED : characteristic of mountain farmsteads

(after Brunskill 1971)

Farmstead, Lancashire (19th.C)

(after Woodforde 1983)

Cumbrian longhouse: cross-passage to house and farm buildings

Quarrymen's cottages, North Wales

Longhouse of granite and thatch, Devon

Cob and thatch in Devon

Traditional two-roomed cottage, rough stone, heather thatch

One and a half storey house of Highlands and Islands

Wheelhouse, Northumberland (19th. C)

Northumbrian defensible bastle

Yorkshire dales, dwelling house with mullioned and transomed windows

East Yorkshire, double-pile house with accommodation for labourers at one end

East Midlands, framework knitter's workshop

West Midlands, hall and crosswing: half framed timber

(after Brunskill 1985)

Cotswold dwelling, stone, stone-tile and dormer

Tall stone Cotswold barn

Wealden house of SE

Ailsed barn, East Anglia

Cotswold cottages, one room and loft

Map labels: Highlands and Islands of Scotland; Central Lowlands; Southern Uplands; North - East England; Cumbria; Yorkshire Dales; North - West England; East Yorkshire and Humberside; North and East Wales; South Pennines; East Midlands; West Midlands; Cotswold and Limestone; East Anglia; South and West Wales; Home Counties; South - West England; South - East England

Figure 4.6

approximate representations of contrasts which can be detected on the ground. The East Yorkshire and Humberside region embraces diverse countrysides. The distinctive rolling lowland of Holderness, where the land fades away into the sea, is followed by the Yorkshire wolds, a bare chalk upland, now ordered and tidy, but ravaged by long centuries of occupation. To the north the great Vale of Pickering still reflects the fact that it was once a great ice-dammed lake in its marshy bottom and sandy edges while further north there are sharp contrasts between the fertile Tabular Hills and the infertile deltaic sandstones of the North Yorkshire Moors. Each of these vernacular regions is in fact made up of distinctive smaller tracts, too small to be depicted in a generalised map, and while regions may have historic substance, i.e. have long been identified – the very name *wold*, having roots in the Anglo-Saxon *wald*, wood – such distinctions are nevertheless divisions of convenience. The vernacular diversity seen in Figure 4.6 touches many essential qualities which give character to settlements. One quality is from the land itself, the red and tawny rock and earth of the Midlands, the golden yellow of the Cotswolds, the russets of the south-eastern timber counties and the greys and whites of the north, but with these geological properties is closely allied a further factor, the fundamental quality of light: softer inland, bright near the sea and often harshly clear in favourable northern lights. A second element comes from the character of the underlying cultural bases and overlays, seen both in the building techniques, starkly clear in these simple illustrations mainly derived from Brunskill's work, and even more clearly in the finer details of decoration and style (Brunskill 1985; Penoyre and Penoyre 1978; Woodforde 1985).

Conclusions

Given the title of the chapter it is necessary to reflect on the extent to which any of these rural buildings provided 'individual space'. The short answer is that most did not. 'A room of your own' is a comparatively modern concept, which has, through many centuries, been filtered down from the great houses of the wealthy. Shared rooms, shared attics, shared beds, shared lives have been and are the common lot of most people, involving a close, noisy, often smelly intimacy between people and livestock. Only in the Victorian period do we find estate cottages being planned with two or more bedrooms, with the avowed intent of separating the sleeping quarters of the unmarried males and females, siblings, lodgers and servants. In Japan, in contrast, there has been, at least until very recently, little concern with privacy at any social level – the house can be seen right through – and if guests stayed the night, all slept next to each other, intermingling sexes, strangers and households alike. In all of the societies discussed, in addition to the master and mistress of the house, husband and wife or wives, there were children and older people, young unmarried adults, the widowed or bereaved, the apprentices and the servants. Of course, 'individual space' is both physical and psychological. It is the area within which elemental rights can be exercised, and the area where respect, personality, awe, fear or love grants avoidance of intrusion.

The buildings described here are settings for the lives of individuals and families, and the underlying harsh need to earn a living is always present, radically affecting the building materials which are used, the scale and plan of each structure, the frequency with which it is extended or rebuilt, the continuity or discontinuity of possession and the underlying arrangement of the settlement forms and patterns of which each individual building is but a part. These forces, in varied disguises, have always been present in rural dwellings. However, once formal architecture has been achieved, the buildings reflect more than these basic needs; they reflect the owners' place in the broader society, their wealth, their status, their aspirations, the interacting complex of concepts and people within which most human beings must live their lives. Even the most basic buildings involve communal effort, communal aspirations and communal attitudes. Traditional buildings represent a continuity of habit and attitude, a meaning encapsulated within the word 'vernacular', but this begs a crucial question: should the settlement forms, themselves associations of private, communal and public space, also be seen as an expression of vernacular traditions and values?

Plates

Only aerial photographs can reveal the qualities of settlement forms and patterns, yet these add an element of distance which often reduces the subject to a map-like structure or even an abstract design. The photographs here are chosen to show the characteristics of particular places as caught through the necessarily narrow view of a camera lens. They also capture the varied scales of enquiry involved in the study of landscapes of settlement as well as the richly varied qualities of the particular places.

Plates 1 and 2 are courtesy the Royal Geographical Society/Caldicott; Plates 3–5 are courtesy Dr W.T.W. Morgan; Plates 6–9 are courtesy the author.

Plate 1 This traditional Bedu goatskin house, *Beit-sha-ar* (literally 'house of hair') characterises a nomadic lifestyle with ancient roots along the desert margins of the Middle East and North Africa, but modernisation is seen in the form of a truck. In fact most of these pastoral farmers have permanent home bases from which they make their extensive migrations to find feed for their flocks

Plate 2 The modern settlement of Safawi, north-east Jordan, is a desert outpost: a few dwellings originally grew up around pumping station H5 on the pipeline of the Iraqi Petroleum Company, but Safawi is now a stopping point for the many trucks and tankers which ply this route. These two elements, a trade-route and a convenient stopping point, are simple causal factors which underlie many long-established settlements. In this case the pumping station was crucial, but more usually and in former centuries, the availability of water, the presence of a river crossing or a portage point, a stronghold where tolls had to be paid, or the presence of a local market, have served to arrest movement for a short time

Plate 3 Three plates (3–5) show African settlements, although it is sobering to reflect that the village of Keyala, in the Equatorial Province of the Sudan, may no longer exist, for it was photographed in 1974. Comprised of materials drawn wholly from the surrounding land, this pastoral settlement comprises a series of living huts forming an agglomerated compound. The steep roofs of reed thatch repel the occasional heavy rainfall of this savanna region, while the dry-stick fences and parched ground reflect the long dry period

Plate 4 A broader view of a similar type of agglomerated village is seen at Wiatua, in the Lafon area of Equatorial Province, Sudan. The settlement comprises a series of adjacent compounds, each separated by communal ways, and with an open space used for protecting cattle and for ceremonial at the centre. The settlement is sited upon a rock outcrop, with the hill mass from which the photograph is taken ensuring a water supply. The variations in hut size reflect social status in this polygamous pastoral society

Plate 5 Sura Village, set on the Jos Plateau of Nigeria, is photographed from the steep slopes of a volcano, Kereng, and forms not a single nucleation, but a series of associated compounds. These occupy the lower slopes of the volcano, where the soils are of better quality, and where water is available because of drainage from the mountain. A permanently cultivated area, subjected to gardening as much as farming, is subdivided by euphorbia hedges. These served the triple purpose of (i) enclosing and delimiting the small fields, (ii) where double they delimit trackways, while (iii) the whole complex had a defensible quality when the community was subjected to slave raiding by horsed Hausa warriors. Beyond lies open land, formerly used as grazing, but in 1976, when the photograph was taken, used for extensive arable

Plate 6 Only careful attention to detail reveals the presence of settlement around this village pond at Besser, Samsø, Denmark. Now an attractive scenic element, such ponds had an historic importance as water supply points for both humans and stock amid the well-drained morainic landscapes which are characteristic of the island. Excavations, in a very dry summer, at a tiny but persistent pond on a small now unoccupied adjacent island produced quantities of neolithic flint implements, suggesting that even a pond may represent an archaeological site of great importance and be a vital component in encouraging the nucleation of settlement amid seasonally dry landscapes

Plate 7 Staindrop, County Durham, England, is a fine example of a regular street green plan. In this case the building line is 'closed', i.e. wholly infilled by eaves-orientated buildings. In fact, slight ambiguity is present, for Staindrop once possessed market rights, and in common with many such large green villages, occupied the threshold between a true village and a true town. Similar plans are common throughout northern England, Scandinavia and Eastern Europe. The antiquity of settlement on or near this site is attested by the church, the nave of which is seventh century in date

Plate 8 This picture shows the excavated interior of the church at Wharram Percy, East Riding, Yorkshire, England. The structural remains and foundations range in date between the early tenth and the nineteenth century, and some elements may be even earlier. They reflect the continuity of this sacred site within the life of a village, continuing even after the village itself was deserted in the sixteenth century

Plate 9 This roadside view of Thoralby, North Riding, Yorkshire, England, captures both the site and the form of this northern village. Built of grey sandstone the houses are aligned along a break of slope: behind, rising, well-drained land is now down to pasture, but once represented the core of the village arable and was maintained in good heart by such manure as was available. Successive intakings carry stone-walled fields upslope, while in the middle ground an outlying small hamlet has arisen by a farmstead. First documented in 1086 as *Turoldsbi*, Thurold's homestead or Thorold's village, the ambiguity of the interpretation of the name echoes the fact that many villages of this type represent decades or even centuries of accretive growth. Now in a National Park, the Yorkshire Dales are now valued for their aesthetic qualities, and tourism has become an important element in the economy, a part of the diversification of farm incomes

CHAPTER 5

Settlement forms

A contradiction permeates all studies of settlement forms. On the one hand is the theoretical impossibility of creating a viable classification of an apparently infinite variety of forms when there is evidently an absence of any direct connection, in either time or space, between most of them. Identical or similar forms can be generated in wholly different ways and at diverse times. On the other hand, it is also evident from the most superficial observations that settlement forms have close and complex relationships with human culture, reflecting lifestyles and aspirations. While there is no doubt that settlement forms can and should be studied as part of the societies generating them – and in this it is interesting to note how much important work concerning them derives from the investigations of anthropologists – there are important questions about their role as cultural indicators when their characteristics are plotted on distribution maps. While the interpretations which identified distinctive settlement forms with racial groups have now been largely abandoned, we are, nevertheless, left with a powerful dilemma. The observable spatial variations in the characteristics of settlement plans continue to beg many questions. These can be seen both in landscapes and on maps, at continental scales (Figures 7.4, 7.5 and 7.6) as well as quite local scales. To use these rich data, classification is an essential step, because in order to talk about rural settlement forms at all there must be a framework for description and analysis, comparison and generalisation. Nevertheless, classification must, above all, not be an end in itself but a tool to aid enquiry and rigorous investigation. In this, the morphological approach can be used as a powerful research tool, complementing, amplifying and extending the limited written record. The volume of evidence is very great. As with the architecture of individual buildings, careful structural analysis of particular cases coupled with careful comparative studies leads to broader hypotheses and questions.

The first part of this chapter will describe a system of classifying settlement forms, followed by an examination of a series of examples drawn from worldwide contexts. Generalisations emerge about the contrasts between structured and unstructured, planned and unplanned settlements. Second, the fact that settlement plan types, once identified and classified, can be observed at different stages of development leads to discussion of the processes of change which affect them. Third, the recognition that not only does change occur within a given type of plan but that individual plan forms do not occupy watertight compartments, and a single farmstead can become a hamlet, a hamlet a village, and so on, implies a need for some assessment of rural settlement dynamics.

Classification: a morphological framework

The system of classifying settlement forms presented here draws upon ideas already examined in Chapter 4 (Figure 4.1i). The framework consists (Figure 5.1) of a series of boxes in a grid, within which individual settlement plans can be placed. However, there is more to the approach than this, for the grid forms one step towards a more complex model designed to accommodate ideas about the processes which affect both individual forms and forms within patterns, changes which happen in both space and time (Roberts 1987a). This more complex version, converting the grid into several layers in a matrix, appears as Figure 5.10, and will be discussed later in the chapter (see p. 115). An understanding of the principles involved in creating and using this classification provides a basis for thinking about settlements, a framework within which further observations and qualifications can be accumulated.

A convenient starting point for beginning to understand how settlement plans can be analysed is seen in Figure 5.1a, where the hailing distance of 150 metres, already noted in Chapter 2, is defined. Where three farmsteads exist in close proximity we have the beginnings of a nucleation, and already at this stage one of two basic shapes will tend to emerge, for the three must either form a row or a triangle. It is, indeed, just about possible to arrange the three in a shape which is neither, but once an individual steading is placed in an enclosure and more dwellings or farmsteads are added to the nucleus, then the basic row/agglomeration division is defined and as growth occurs one or the other tends to be sustained and reinforced. At this point, depending upon the way the individual farmsteads are arranged relative to each other and the degree to which any associated enclosures are structured, a division appears between regular and irregular layouts. There is the added complication that agglomerations may be either rectangular, based on a grid, stellate, or even circular; this is equally true of both regular and irregular agglomerations. Finally, nucleations may or may not contain an interior open space, a green or plaza. This allows the grid seen in Figure 5.1d to be structured (it is of course split by the page division). With one exception the cases included here are drawn from England, but as the examples used in Figure 5.2 show, the system has a worldwide application (Uhlig and Lienau 1972).

At this stage of discussion this classification creates a convenient way of thinking about the shapes made by the plans of hamlets, villages and small market towns in the landscape or on maps, but there is an underlying question. One example will show the problem: taking the three farmsteads of the simplest clusters, these may either operate wholly individually and independently, with each family working their own fields, or in contrast the three families may share many communal resources, particularly arable lands, meadowlands and grazing lands. A mere plan is a poor surrogate for understanding the functional dynamics of a settlement, yet it is often the only source. In general, as the numbers of dwellings in a nucleation increases, communality, already discussed in Chapter 2, increases, binding the inhabitants together into a community of effort. This can override even marked differences in wealth, social status or lifestyle, for the sum is greater than the parts.

Nevertheless, this grid has many uses, and two are of particular importance: first, it helps define varieties of plan types and allows the formulation of a terminology to describe and compare; thus Appleton-le-Moors is a regular street village, while Cardington is an irregular agglomeration, neither possessing greens. Middridge is a regular green village while Walbottle is an irregular agglomeration with a green, and so on. There is no need here to list the formal description of each and every plan type present, for appropriate terms can be worked out with a little common sense, but the grid structures any descriptions and provides a framework for reference, allowing comparison, discussion and analysis. Second, associated with the drawing of each example is a small symbol, an ideogram, and by using these it is possible, with some precision, to map the character of village plans over large areas. This procedure may, for example, be compared to soil mapping, where a series of careful studies of sample profiles are excavated and studied in individual pits, and then this knowledge extended, using an awareness of geology, slope, the limited evidence of auguring and further test pits, to map a region. There are also parallels with the technique of recording geomorphological features by using symbols. The preparation and analysis of such maps is a rather specialist task, and the end-product is apt to resemble a musical score, interpretable by those who have taken the trouble to grasp the meaning of the symbols.

Figure 5.2 compresses the same argument into another diagram, but in this case, because it uses examples from widely dispersed locations, shows that the classification system does indeed extend beyond Europe. In fact the grid represents more than a set of posting boxes and to appreciate its wider potential a number of points must be made. First, the plans used in the diagrams are all to the same scale. For the purposes of comparison this is essential, so that the dimensions of Moshav Nahalal founded in 1921 (Figure 5.2) may be compared directly with the settlement at Walbottle, Northumberland, England, recorded on a plan of about 1620 (Figure 5.1). Scale comparison is a fundamental requirement in morphological studies, revealing what is normal, what is big and what is small. Second, each of the cells in Figure 5.1d defines a basic plan type, i.e. a distinctive arrangement of private, communal and public space (Figure 4.1), but if a whole series of plans are studied at a scale of 1:10,000, approximately 6 inches to the mile (a minimum for such work), then settlements will quickly be found that do not fit easily into any cell. The implications of this idea can be most readily grasped by looking at Appleton-le-Moors; if an earthquake were to tear the village in half, with the crack running exactly down the middle of the street, then a point would be reached when sufficient open space is created to be termed a 'green', and eventually a strip about 100 metres wide could appear, the width of the green at Middridge (Roberts 1987a: Chapter 2). The point is that in the real world cases can be found at all points along the conceptual scale extending from a Middridge to an Appleton type of plan. To borrow a word from soil science we have defined a *catena* of types. Of course, for mapping purposes a decision must be taken to allocate a given settlement to a cell, but in reality catenas exist across all axes of the grid, allowing the system to accommodate a great range of possibilities. In practice this does not mean that the classification dissolves in an unmanageable morass of types, it means that the classification is broader and more subtle than it may at first appear. Of course, the grid cannot directly answer the most fundamental question of all: why do the processes associated with the generation of nucleated settlements result in this range of plan types?

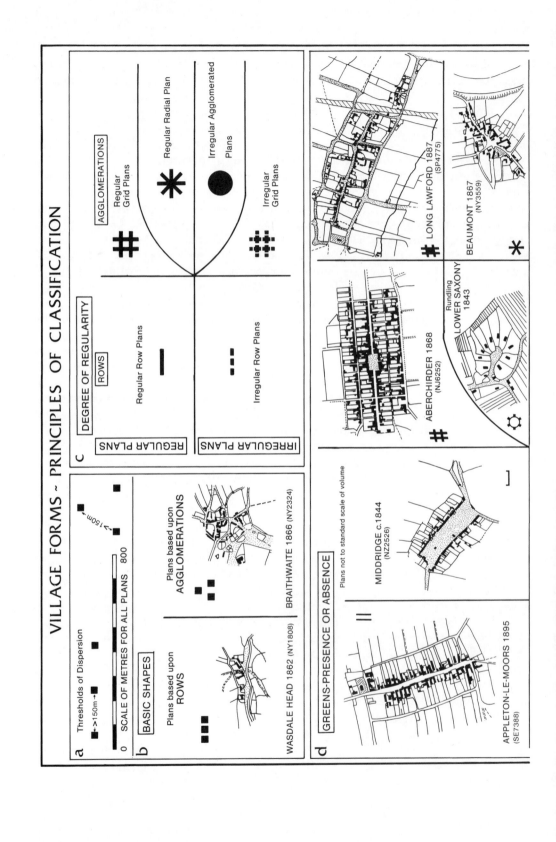

VILLAGE FORMS ~ PRINCIPLES OF CLASSIFICATION

a
Thresholds of Dispersion

■ >150m
←150m→

SCALE OF METRES FOR ALL PLANS
0   800

b BASIC SHAPES

Plans based upon ROWS

Plans based upon AGGLOMERATIONS

WASDALE HEAD 1862 (NY1808)

BRAITHWAITE 1866 (NY2324)

c DEGREE OF REGULARITY

REGULAR PLANS

IRREGULAR PLANS

ROWS

Regular Row Plans

Irregular Row Plans

AGGLOMERATIONS

Regular Grid Plans

Regular Radial Plan

Irregular Agglomerated Plans

Irregular Grid Plans

LONG LAWFORD 1887 (SP4775)

BEAUMONT 1867 (NY3559)

ABERCHIRDER 1868 (NJ6252)

Rundling LOWER SAXONY 1843

d GREENS-PRESENCE OR ABSENCE

Plans not to standard scale of volume

MIDDRIDGE c.1844 (NZ2526)

APPLETON-LE-MOORS 1895 (SE7388)

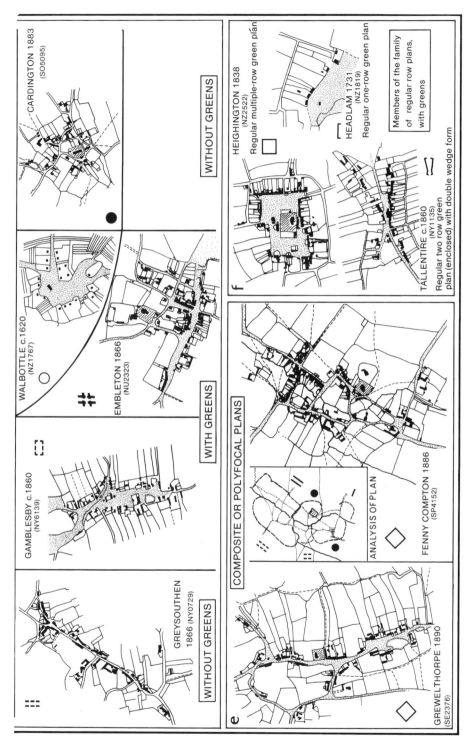

WITHOUT GREENS

CARDINGTON 1883
(SO5995)

WALBOTTLE c.1620
(NZ1767)

EMBLETON 1866
(NU2323)

WITH GREENS

GAMBLESBY c.1860
(NY6139)

GREYSOUTHEN
1866 (NY0729)

WITHOUT GREENS

HEIGHINGTON 1838
(NZ2522)
Regular multiple-row green plán

HEADLAM 1731
(NZ1819)
Regular one-row green plan

Members of the family
of regular row plans,
with greens

TALLENTIRE c.1860
(NY1135)
Regular two row green
plan (enclosed) with double wedge form

f

COMPOSITE OR POLYFOCAL PLANS

ANALYSIS OF PLAN

FENNY COMPTON 1886
(SP4152)

GREWELTHORPE 1890
(SE2376)

e

Figure 5.1

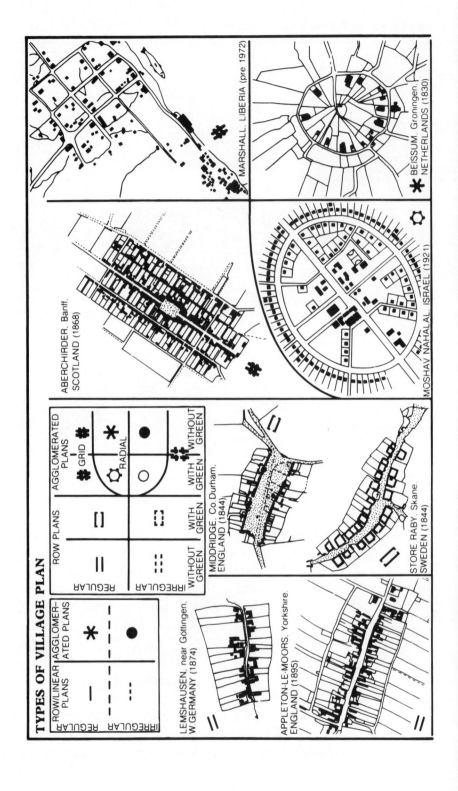

TYPES OF VILLAGE PLAN

ROW/LINEAR | AGGLOMER-
PLANS | ATED PLANS

REGULAR | — | ●
IRREGULAR | ⫶ | ⫶

LEMSHAUSEN, near Göttingen,
W.GERMANY (1874)

APPLETON-LE-MOORS, Yorkshire,
ENGLAND (1895)

	ROW PLANS		AGGLOMERATED PLANS
	WITHOUT GREEN	WITH GREEN	GRID	RADIAL
REGULAR	‖	[]	✳ GRID	● WITHOUT GREEN
IRREGULAR	⫶	[⁖]	✿ WITH GREEN	○

MIDDRIDGE, Co. Durham,
ENGLAND (1844)

STORE RABY, Skane,
SWEDEN (1844)

ABERCHIRDER, Banff,
SCOTLAND (1868)

MOSHAV NAHALAL, ISRAEL (1921)

MARSHALL, LIBERIA (pre 1972)

BEISSUM, Groningen,
NETHERLANDS (1830)

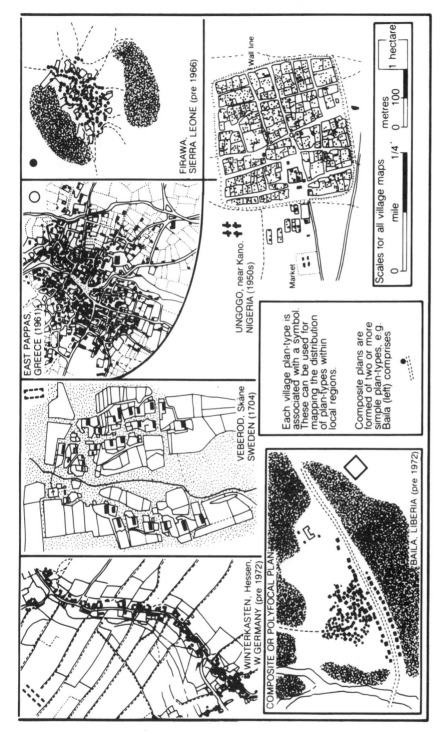

FIRAWA, SIERRA LEONE (pre 1966)

EAST PAPPAS, GREECE (1961)

VEBEROD, Skåne, SWEDEN (1704)

WINTERKASTEN, Hessen, W GERMANY (pre 1972)

UNGOGO, near Kano, NIGERIA (1950s)

Wall line

Market

Scales for all village maps

0	mile
1/4	
	1 hectare

metres
0   100

Each village plan-type is associated with a symbol. These can be used for mapping the distribution of plan-types within local regions.

Composite plans are formed of two or more simple plan-types, e.g. Baila (left) comprises

COMPOSITE OR POLYFOCAL PLAN

BAILA, LIBERIA (pre 1972)

Figure 5.2

Third, as Figure 5.1d shows, Middridge is a type of plan built of two compartments; Headlam comprises one, Heighington at least three, probably four, while Tallentire has a distinctive dumb-bell shape. All of these plans are in fact related to those seen at Appleton and Middridge, and can be accommodated more easily in a matrix (Figure 5.10) than in the simple flat grid – in fact, as will be shown below, the matrix embraces all settlement types, from a single farmstead to a market town. Next, many villages comprise not just one plan type but several, i.e. they are composite in character. This is suggested in the case of Grewelthorpe and Fenny Compton (Figure 5.1e), but is even more clear in the case of Baila (Figure 5.2, lower box), where growth along a roadway is drastically altering a traditional African agglomeration made up of numerous compounds (see Firawa, Figure 5.2). Finally, it is essential to stress that this system of classification is not bound by time or by space; any settlements of any period and any place may be related to it.

So far this discussion has done no more than hint that the classification system represented by the grid can be painlessly extended up the hierarchy to market towns: it can indeed be extended downwards to assist understanding of dispersed settlement forms. It is, however, wholly fair to ask 'What do the cases included in Figures 5.1 and 5.2 really tell us?' Irrespective of the details of each place, the examples given suggest that some settlements grow in an apparently unstructured way, while others are carefully and meticulously planned. This is an important observation, for careful study shows that even the 'irregular' plans may contain elements of spatial order, while the questions 'Is order present?', 'Why was it created?', 'When was it done?', 'What is its future?' lead far beyond the confines of a short study.

In the discussion of Figures 5.3 and 5.4 two themes will be pursued, the contrasts between regular and irregular types and questions of change and stability in plans.

Regular and irregular plan types

The plans appearing in Figures 5.1 and 5.2 may be compared with those shown in Figures 5.3 and 5.4, where the examples selected range from Europe to India and from Africa to Vietnam. Once again, all are brought to the same scale, a task not without problems as Phuong Vy and Old Kwamo show. To create these diagrams a measure of standardisation has been imposed. The word 'regular' when applied to settlement plan suggests the presence of systematic structure or order, to be seen in the arrangement of buildings and/or plots. There is a high probability that geometrically regular settlement forms derive from a planning process, however this was generated or imposed. Nevertheless, the converse is not necessarily true, for while vast numbers of settlements are visually chaotic, lacking any obvious structure or order, in the eyes of their inhabitants a sub-surface or deep structural order may be present, only discernible by social anthropologists after prolonged investigation. For example, in Africa traditional settlement forms have been noted whose arrangement conceals a layout thought to reflect the shape of a turtle or the parts of the human body. This is a perception of the built environment which sees it as a reflection of a divine model, a cosmological frame, ultimately linking buildings, settlement, indeed the whole 'architecture' of the territory occupied by a group, to their perception of the wider ordering of the universe.

There is, however, a more general point here. If a settlement is indeed planned, then

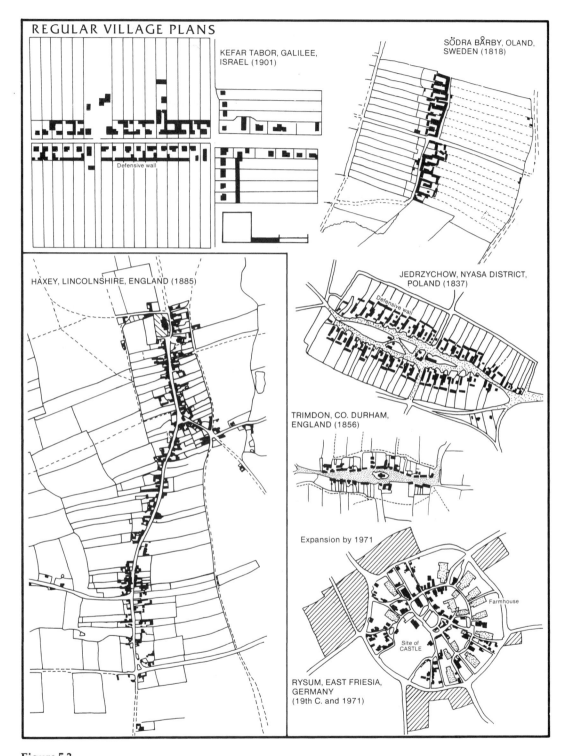

REGULAR VILLAGE PLANS

KEFAR TABOR, GALILEE, ISRAEL (1901)

Defensive wall

SÖDRA BÅRBY, OLAND, SWEDEN (1818)

HAXEY, LINCOLNSHIRE, ENGLAND (1885)

JEDRZYCHOW, NYASA DISTRICT, POLAND (1837)

Defensive wall

TRIMDON, CO. DURHAM, ENGLAND (1856)

Expansion by 1971

Farmhouse

Site of CASTLE

RYSUM, EAST FRIESIA, GERMANY (19th C. and 1971)

Figure 5.3

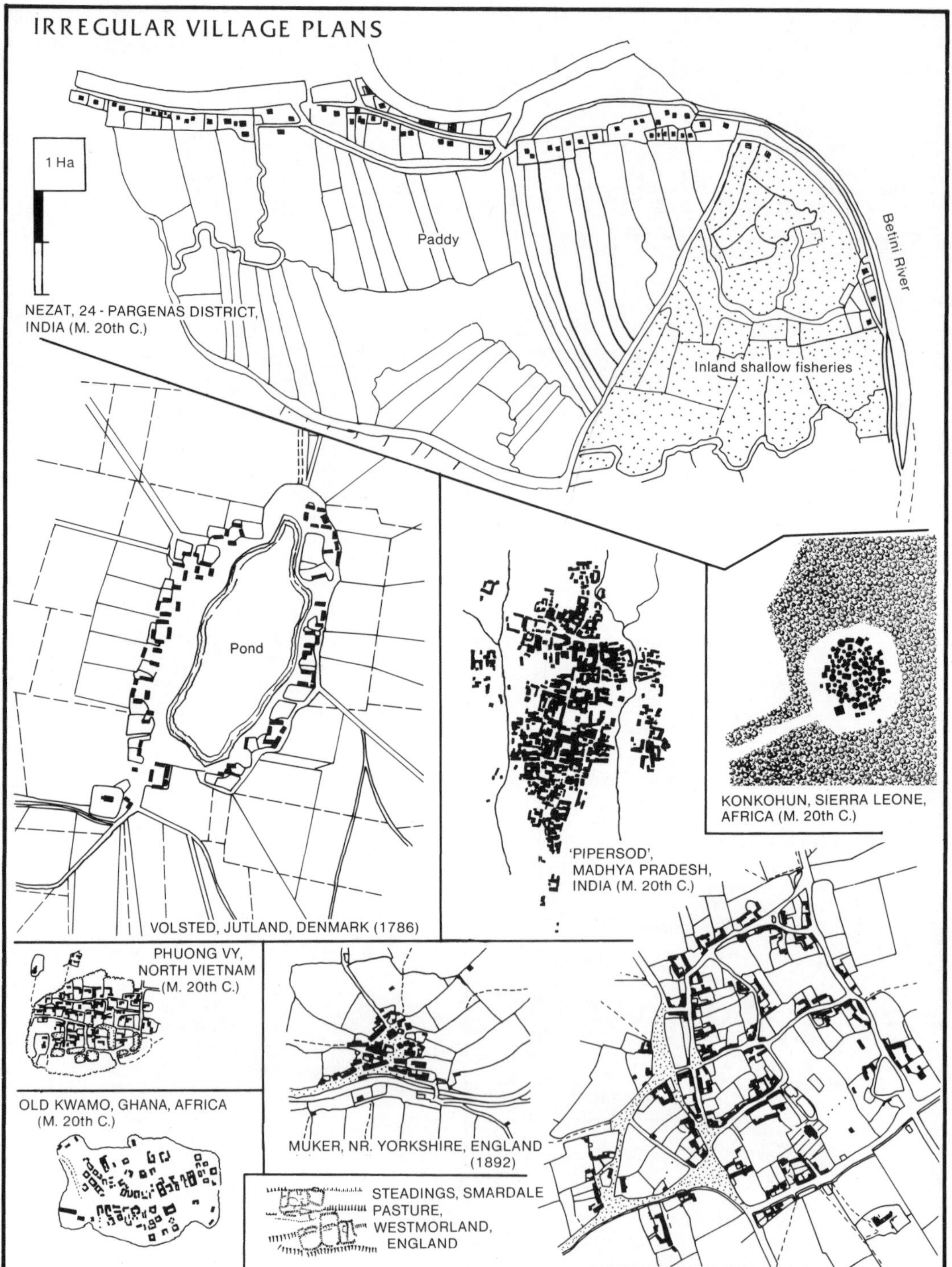

IRREGULAR VILLAGE PLANS

1 Ha

NEZAT, 24 - PARGENAS DISTRICT, INDIA (M. 20th C.)

Paddy

Betini River

Inland shallow fisheries

Pond

VOLSTED, JUTLAND, DENMARK (1786)

'PIPERSOD', MADHYA PRADESH, INDIA (M. 20th C.)

KONKOHUN, SIERRA LEONE, AFRICA (M. 20th C.)

PHUONG VY, NORTH VIETNAM (M. 20th C.)

OLD KWAMO, GHANA, AFRICA (M. 20th C.)

MUKER, NR. YORKSHIRE, ENGLAND (1892)

STEADINGS, SMARDALE PASTURE, WESTMORLAND, ENGLAND

KILSBY, NORTHAMPTON, ENGLAND (1884)

Figure 5.4

the people who created it had in their mind's eye an image, a concept, of what a settlement should be. This is exciting, for not only are these images derived from roots deep within cultures and lifestyles, they can travel as a package in the mind, be elaborated or simplified, duplicated exactly or adapted to changed or local circumstances, and Figures 5.1 and 5.2 are testimonials to the fertile imagination of human beings. To take one specific example, Kefar Tabor (Figure 5.3) was founded in 1901 in Palestine by Jewish refugees from Russia, mainly professional men, merchants or artisans (Blake 1969); each family was allocated about 15 hectares of land in private ownership, and cereal production was the mainstay of the economy. The layout of the village, with small rectangular plots attached to the farmsteads, was ultimately derived from the grain-growing villages of southern Russia, while the line of buildings at the rear of the farmsteads, seen by Blake as defensive walls, can be closely paralleled throughout many parts of eastern and western Europe – in short, the origin of this type of plan is a historical problem, and has little to do with the colonising context in which it was established in this particular case. Kefar Tabor may be compared with the much more imaginative and sophisticated layout of Moshav Nahalal in the Yezreel Valley (Figure 5.2), planted in 1921 and designed by the architect Richard Kauffmann, who drew together seventy-five households in a mixed farming economy, with the intention of fostering a close-knit community spirit; concrete cowsheds on the settlement's perimeter afforded a defensive ring, echoing an older tradition (see Figure 5.3, Jedrzychow). The sheer concentration of activity within the key ring, dwellings, byres and stores, movements in and out to the fields to cultivate and harvest, eventually placed great demands on plots with a shared entrance. Each frontage was only 22 metres, with farmyards only 30 metres wide, and with only 40 metres between the farmhouse and the cowshed. We have here an interesting glimpse of the way technological advance can generate the need for change within long-lived plan types, a factor which in other contexts (Figure 6.8) may account for the willingness of farmsteads to migrate from the village core to new sites amid ring-fenced fields.

Also included in Figure 5.3 are Trimdon, Jedrzychow, Haxey and Södra Bårby, which may be compared with Middridge, Store Råby, Appleton-le-Moors and Lemshausen in Figure 5.2; all of these plans show broad structural similarities: the focus upon an axial street or green, geometrically defined compartments of house/field plots and the presence of an interior green and back lane. In western Europe all of these have roots which are medieval. Jedrzychow is a type Polish scholars link with regular medieval layouts in which cords or ropes were used to survey the site, while Trimdon and Middridge are representative of types found throughout the north of England whose regular compartments, orientated north–south or more commonly east–west, are documented in contexts before 1200, possibly even three-quarters of a century earlier (Szulc 1968; Homans 1941: 83–106). The east–west orientation, sometimes clearly site-controlled, sometimes not, is associated with cosmological beliefs which divide the village and its fields into two halves, a sun side and a shadow side, beliefs which in Scandinavia appear as a formalised system of settlement planning and field division known as *solskifte* (Roberts 1987a: 3.7). It is probable that many hundreds of villages and hamlets throughout northern Europe have plans incorporating these ideas, a mixture of the practicalities of dividing the arable into strips and underlying cosmological beliefs. There are, of course, many complications, for villages have been destroyed and rebuilt, replanned and moved, while most of their buildings have been replaced as the natural forces of decay demanded and as economic circumstances allowed. Nevertheless, this type of plan, still a living part of many landscapes, closely reflects medieval antecedents.

The ultimate origin of such European regular plans and their associated communal arable fields remains an open question. A type of planned village based upon regular compartments and dating from the tenth century appears in the northern Netherlands, while small planned villages are known from Denmark as early as the fourth and fifth centuries of this era. Although there is no known lineal linkage, the planning of Kolomiischina (Figure 2.2) reveals that human societies have imposed formal layouts upon settlements from the earliest times (Piggott 1965: 55; Champion *et al.* 1984: 133). It is likely, indeed probable, that the concept of an 'ideal' village was developed at various times in different societies, and we need not postulate diffusion from one point of origin. The concepts underlying the similar plans found throughout Europe emerged as part of an intellectual heritage which may ultimately have roots in the ordered world of Rome, where the standardisation of the layout of military camps, although not absolutely uniform, was substantial, and one source, perhaps, for the idea of small, standardised layouts (Johnson 1983). Furthermore, there are indications, in the form of village fiscal assessments which are then sub-divided in equal proportions among individual farms, that an underlying motive for this imposed regularity was to share the burden of taxes among the farmers. On the other hand, as settlements increased in size and as resources became more stretched, the farmers themselves must have felt the need for ways of defining not only the location of their dispersed arable strips, but the associated shares in the meadows, the woods and pastures, the turf moors. Indeed an ordered village plan could be used as a paradigm for sub-dividing any conceivable environmental resource in this broad northern European region once wholly dominated by mixed farming based upon wheat, barley, rye, oats, and the breeding of cattle, sheep, pigs and goats. In wholly regulated villages, where the regular plan was linked to a system of sharing burdens and allocating resources, the house plot, or *toft*, of a standard size became, as the Danish proverb states 'the mother of the acre' (i.e. the arable lands), so that a half toft carried half shares, a double toft two shares and so on. Deep reverberations from these ancient European arrangements are undoubtedly present in the idealised layout at Kefar Tabor and reach their ultimate expression in the plan of Moshav Nahalal.

The case of Haxey raises other questions; it shows quite clearly that the settlement originated as a small regular plan, visible as a tight rectangle delimited by former back lanes now reduced to footpaths. By the later years of the nineteenth century, it had expanded over the former arable strips of its own fields, the tight core and ribbon of extension being clearly seen in the map of 1885. To pursue these ideas further would carry the argument deep into the realms of historical geography, not the brief of this study, but the point must be made that some grasp of historical circumstances is essential if the traditional rural villages of Europe are to be appreciated and understood, for the extension of Haxey is one aspect of change which is still affecting settlement forms inherited from past centuries. At times the processes of change are very gradual and at other times they are overwhelming and catastrophic, but in many thousands of cases the 'past in the present' is still detectable to the discerning eye.

The final example in Figure 5.3 is of a regular radial plan at Rysum, northern Germany (Mayhew 1973: 32). Here the underlying regularity is radial, and another example, Beissum in the Netherlands, is found in Figure 5.2 (Harten and Schuyf in Roberts and Glasscock 1983: 43–59). These distinctive settlements, built in response to a gradually rising sea level, are placed on artificial islands raised a metre or so above the surrounding coastal marshlands of the north European coasts. Rysum is a rather large example and its summit reached no less than 6 metres above the surrounding marshes. These mounds, *Terpen* or *Wurten*, are almost wholly artificial, being constructed of brushwood, hurdles, animal manure and other cultural debris as well as mineral soil. They seem to have originated in the first centuries AD and were progressively added to. Occupation in general ended in the fifth and sixth centuries and many were destroyed during the nineteenth century for they formed a rich source of fertiliser to spread upon the marshlands then being reclaimed by vast dykes. Nevertheless, a proportion are still in use, and Rysum is a splendid example. It is probable that the double ring roads represent growth phases. The settlement's interior, in this late phase now filled with small cottages housing craftsmen and farm workers, was once more open, with a central green area where stock could be driven in times of flood. Beyond this core lies an almost complete ring of large radially placed farmhouses, the generally regular arrangement, in this case, deriving from the distinctive site. These great East Friesian farmsteads concentrate under one roof the dwelling and all of the essential farm offices, the interior space supported by four vast posts being reserved for hay (Figure 4.4b). Like many vernacular buildings they are now often adapted and in measure modernised, yet preserve some older structural components and overall plan. Recent expansion is altering the ancient form, and additions during the 1960s and 1970s are carrying dwellings out onto the former marsh because modern dykes and pumping technology have made this landscape seem less hazardous than it was in earlier centuries.

To move from ordered regularity to apparent disorder, the case of Nezat (Figure 5.4) shows only a small part of the great chain of homesteads extending for some 5 kilometres along the levees and dykes which fringe one of the numerous rivers and streams of the Sunderban areas of the Ganges delta (Chatterjee and Saha in Singh 1972: 416–21). Nezat village territory covers a total of over 810 hectares (nearly 2,000 acres), the size of a normal parish in the Midlands of England, and in 1961 it included 296 homesteads. Until the 1930s the Sunderbans were a forested region and were colonised when landlords brought wage labourers from other regions to clear the woodlands. These, attracted by cheap land, food and shelter, began the process of permanent settlement. The bulk of the population are engaged in agriculture, the farmsteads comprising a small mud-walled enclosure, with one living room, one small kitchen to one side, and often with one or two families sharing the limited space. Courtyard, kitchen garden and small pond complete the steading. The pattern is conditioned by the fact that settlements, along with the roads, are normally located on fractionally higher ground, strung out along the slightly elevated levee banks and man-made dykes between the rivers and the *padi* fields. Because the rivers are tidal, salinity is a problem, permitting only one crop of rice a year during the monsoon season; there is a small amount of fishing. Superficially these settlements are irregular; indeed the unstable nature of the environment poses a fundamental cultural problem, for river floods fed by the Ganges and the Brahmaputra and tidal waves brought from the south by tropical cyclones are recurrent and at times utterly devastating features of life. Nevertheless, a careful study of the map reveals that the strings or ribbons of homesteads are made up of a series of individual cells, some of which are remarkably regular in layout (Figure 5.4). This may reflect growth or recovery phases (i.e. recovery after a disaster) and, more probably, caste differences, although this is not closely recorded in the study from which the map is drawn.

East Pappas (Figure 5.2) is a very different type of settlement plan, an irregular agglomeration, a structureless tangle of lanes and buildings, the latter often being physically close and sometimes touching, a cultural feature generally not found in the pre-modern rural settlements of central and northern Europe. Such marvellous aggregations of dwellings, apparently without formal structure, are a characteristic of the regions around the Mediterranean, extending into the Middle East and beyond (Figure 5.8). High courtyard walls and forbidding doors, inward looking, give privacy and seclusion amid public alleys where humanity and animals swarm (Schwartzberg 1978: 131). The farmsteads of these warm temperate and even sub-tropical regions are traditionally supported by mixed farming based on wheat and barley, the vine, the olive, the date palm, and sheep and goats. It is tempting to speculate if the generously spaced northern farmsteads with their large buildings were in origin a concession to cattle-rearing and the need to reduce losses of very young calves, whose mothers were, in effect, brought into the home and stall-fed, a tradition for which there is archaeological evidence in Europe as far back as the neolithic period. On the other hand, most of northern Europe is safe from the often acute summer drought of the Mediterranean zone and adjacent areas; in this context water supply, though not to be ignored, was by no means the dominant nucleating factor.

Konkohun is a traditional African forest settlement form, constructed earlier in the twentieth century in response to great political instability (Siddle in Clarke 1966: 62–3). Once surrounded by a stockade, the village is also defended by a single track approach through thick forest. In contrast Baila, Liberia (Figure 5.2), is an example of the changes which can be brought by access to better communication, with the addition of a wholly new plan type resulting in a composite or poly-focal plan. Both Konkohun and Baila present striking contrasts to another traditional African form, Old Kwamo, a typical Ashanti village, consisting of a principal but irregular street, in which are found the buildings of social significance, the Chief's Palace, the courtyard houses of the clan elders, the royal cemetery, plus an open space for funerals and one or two shops. The older houses cluster around this, with newer ones being added to the periphery in rough groupings. A network of passageways links the different compounds, allowing movement between them to sustain social and other linkages, while other paths lead outwards to the surrounding lands and farms. The remaining land within the village fence is used for tree crops, shade trees, fruit production and gardens for staple produce. It is important to grasp that the main street may be used as a road, but it is, given the climatic advantages, primarily a place of public display, where chiefs were formerly borne in palanquins, where surplus food was exposed for sale, and where communal activities, now including film shows and political meetings, could take place (Oliver 1971: 162–3). The apparent irregularity of the layout of this tiny settlement fails to reveal underlying socio-economic forces present. This conclusion must also apply to the tiny hamlet of Phuong Vy in the Red River Delta, Vietnam, where similar accretive growth is likely (Robinson 1984: 51).

The problems of inference from plans can be illustrated by the English cases of Kilsby and Muker (Roberts 1987a: Figures 4.2, 4.4), the former a large irregular agglomeration with some traces of interior greens, and the second a tiny hamlet with the dwellings as much placed *upon* the communal open space as surrounding it. There are, in the absence of early map evidence, fundamental questions about their origins. Muker may represent little more than a rather late congregation of cottages and houses – not farmsteads – set on open land near a church, a late-arriving nucleation in a landscape dominated by scattered farmsteads, while Kilsby probably originated in a loose scatter of farmsteads or small hamlets – either a linked farmstead cluster or a linked hamlet cluster.

To give one example, it is possible that the type of green village seen at Volsted (Figure 5.4; Hastrup 1964: 196–7) is a surviving example of a type ancestral to the village of Nordby (Figure 5.5): there is an element of intuition in this deduction, but as the discussion of Nordby will show there is evidence that the buildings in the central portions of Nordby have a different tenurial status to those on the periphery while a field map of the site done by this author in 1982 shows that the present pond occupies the head of a shallow valley and could have once been larger – village ponds are notoriously susceptible to being infilled. Finally, Hansen has traced in great detail the historical growth of a similar Danish village, from medieval times to the present, demonstrating at first the accretion of full farmsteads followed by the sub-division of some, and then the eventual arrival of smallholders to create a complex agglomerated nucleation (Hansen 1959). Of course, without extensive excavation there can be no certain proof, but the progression is likely, and in this case a glimpse of an important general process is obtained.

The processes changing plans

The upper part of Figure 5.5 draws together into one diagram a summary of the varied processes changing settlement plans through time. Basically there are three key elements, *stability, expansion* and *contraction*. Of course, settlements may remain essentially stable for many decades or even centuries. This stability is difficult to explain, but remoteness, conservative attitudes and conservative lordship are possible factors. Expansion takes place in a variety of ways, involving either the sub-division of the existing plots or tofts, the infilling of existing open spaces within a plan or the addition of new plots, usually encroaching upon former arable fields on the edge of the village. It is feasible to sub-divide these processes further, and the list given in the diagram is not exhaustive. A third process, contraction, may involve either shrinkage, or the extreme cases of settlement desertion, sometimes the result of total destruction, sometimes because of a move to a new site. The physical destruction of some buildings and the use of their plots in other ways is common, and is a process associated with partial depopulation or shrinkage. This may result from the out-migration of some of the population or from the effects of plague, or it may be enforced by a landlord, who decides to amalgamate the existing farms into larger units, or change the land-use radically, from arable to pasture, so needing less labour. Nevertheless, this complex diagram, which includes a very large number of possible changes, some quite small, others large-scale and even catastrophic, provides a useful way of categorising what can be seen in the field and on the map. Multiplied many times the changes seen here categorise the details of the larger-scale processes which alter and restructure complete patterns of settlement.

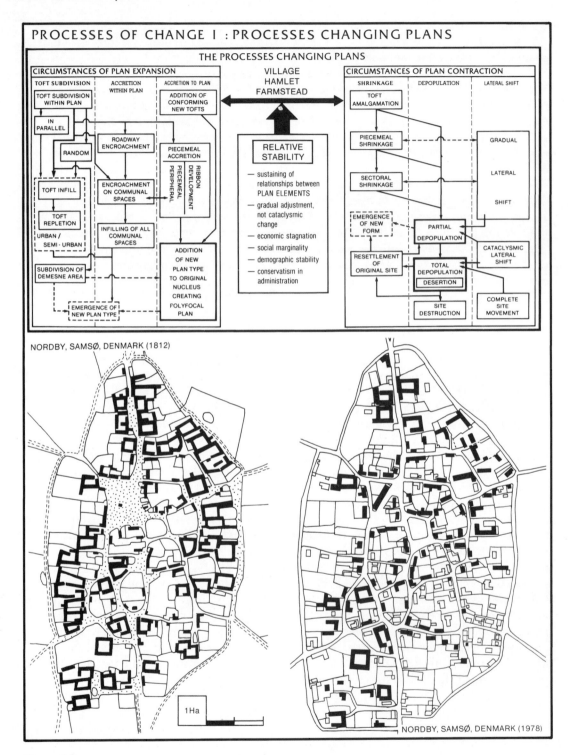

PROCESSES OF CHANGE I : PROCESSES CHANGING PLANS

THE PROCESSES CHANGING PLANS

NORDBY, SAMSØ, DENMARK (1812)

NORDBY, SAMSØ, DENMARK (1978)

1Ha

Figure 5.5

Two European examples of change and stability

The German village of Kiebingen (Figure 5.6) lies on a lower terrace of the River Neckar in Baden-Württemburg; the maps show the settlement at two different stages of development, first in 1819, when it had a population of 581, and was already the end-product of a long and complex development. Superficially it is an irregular agglomeration, but careful study of the plan suggests it originated in one or more compartments of farmsteads, with infill taking place within former green and expansion appearing at the edges. Agricultural buildings still form the heart of the present settlement, but have been adapted, improved and extended. Between 1819 and 1939 the population varied a little, generally falling to between 600 and 700, but after the Second World War some growth took place, with new dwellings added to the periphery, creating a population of 875 by 1950.

Between 1950 and 1985, in this prosperous and attractive part of the state of West Germany, the population virtually doubled from 875 to 1,571. Paradoxically, it was older and cheaper dwellings at the heart of the settlement which were attractive to foreign workers, *Gastarbeiter*, so that now nearly 8 per cent of the population is made up of Italians, Spanish and Portuguese, Turks, Yugoslavs and Greeks, giving a cosmopolitan core to a traditional German village. This is starkly different to what has happened in many parts of England during the same period, where the older buildings in village cores have been seized by wealthy incomers to create their own urbanised image of a rural idyll (Grees 1986).

The chronology of growth phases in Kiebingen is summarised at the bottom left of Figure 5.6, suggesting the earlier elements which may make up the core of 1819 and then showing a more precise chronology of the appearance of the outlying estates. Within car distance from several prosperous small towns, with jobs available in both manufacturing and services, the future of the settlement is assured, but it has already reached a critical threshold: with school, sports and leisure facilities already present, continued expansion will seriously undermine the village's historic character. It has been the subject of a very detailed study, in which geographers, planners and the local population have generated a village plan, a framework for future policy decisions.

The microcosm of Kiebingen is part of a general trend affecting many thousands of villages throughout central and northern Europe. This process can be grasped more readily by consulting the simple model in the lower right-hand part of Figure 5.6: many present villages have origins at least 1,000 years old, and these are represented by the 'origins' stage, comprising a scatter of farmsteads together with a higher-status dwelling of a richer farmer or lord, to which Christianity brought a church. A second stage involves the appearance of a larger, more ordered village, either under the control of the lord or created by the farmers themselves, in which elements of regularity and planning may be present. Throughout most of their history their inhabitants were farmers, subject to the vagaries of weather and farming economics, the whim of the landlord, and the hazards of warfare, but by the nineteenth century population increases, technological advance and rural industries supported more diverse populations, including landless labourers, artisans, weavers, shoemakers, cabinetmakers, schoolteachers, carters and the like. These are the villages whose echoes survive within the living memories of the grandparents and parents of the present generation, although the accessibility of this underlying rurality varies from location to location. It is deeply buried throughout much

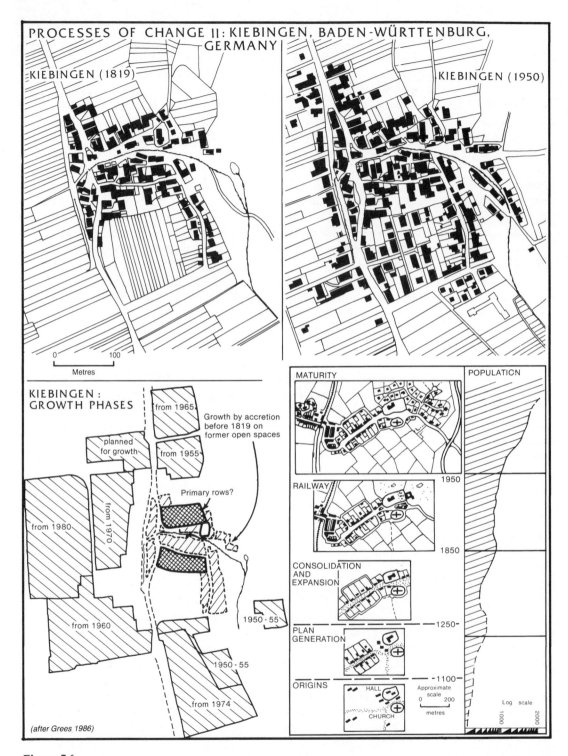

PROCESSES OF CHANGE II: KIEBINGEN, BADEN-WÜRTTENBURG, GERMANY

KIEBINGEN (1819)

KIEBINGEN (1950)

0   100
Metres

KIEBINGEN:
GROWTH PHASES

from 1965

Growth by accretion
before 1819 on
former open spaces

planned
for growth

from 1955

from 1980

from 1970

Primary rows?

from 1960

1950 - 55

1950 - 55

from 1974

(after Grees, 1986)

MATURITY

POPULATION

1950

RAILWAY

1850

CONSOLIDATION
AND
EXPANSION

1250-

PLAN
GENERATION

-1100-

ORIGINS

HALL

CHURCH

Approximate
scale
0   200
metres

Log   scale

1000   2000

Figure 5.6

of England, close to the surface in Scandinavia, Germany and France, and still present in those parts of Europe currently emerging from Communist darkness to the uncertain sunlight of capitalism. It is in the favoured economic zones of northern Europe, where industrial and urban prosperity are generating wealth, that the characteristics of the final stages of expansion and urbanisation of rural villages are found, initially under the influence of the railways and more recently the motor car, although in the nature of things the car now affects many more villages and hamlets than could the railways. The general parallels with the earlier arguments concerning village development in Denmark are not accidental, for while details and chronologies may vary, the broad trends are common to many regions.

This general model can be applied to thousands of particular places and can give an understanding of their development: it is a useful general way of thinking, but it also raises questions. Expansion has frequently been balanced by contraction, caused by disease, warfare or change in the economy, indeed there have been times when whole groups of settlements have been depopulated. Change may be seen as pulses of energy, derived from external economic forces, coursing through the life of rural localities and stimulating growth or decline. A most interesting question concerns the spatial distribution of these pulses: how many settlements are affected and with what intensity? Kiebingen is very characteristic of the Neckar Valley, but would not be typical of what is happening in other regions.

As already noted, one of the strange things about Kiebingen, to English eyes, is that immigrant workers have been able to purchase older dwellings in the heart of the village. In England it is these which are normally the most valued, where those with funds are able to purchase, adapt, modify and 'preserve' older features, albeit within a dwelling which incorporates all the modern conveniences. However, in regions where historic circumstances have bequeathed to the present high-quality vernacular buildings, the result of earlier prosperity or the availability of good local building stone, then the core areas of old villages may now be subject to careful conservation and preservation. Kiebingen is moving in this direction, but the Danish village of Nordby – a large irregular agglomeration – is a fine example (Figure 5.5). The map of 1812 provides a datum, for there has subsequently been almost no expansion beyond the surrounding back lane delimiting the settlement at that date (Nationalmuseet 1978). The village is a classically Danish irregular agglomeration around a green, and probably originated in a more open arrangement such as that seen at Volsted (Figure 5.4). The settlement probably began as a rather irregular clustering of farmsteads near or around a pond, a vital source of water for stock during the drier periods on the rather freely draining morainic soils of Jutland and the Danish islands. Encroachment, the deliberate siting of new buildings on the open green, and eventual infill of the remaining sections with smallholders' and artisans' cottages (compare the map of 1812 with that of 1978), accounts for the present agglomerated form. This is not pure hypothesis, for when the dwellings of the full-status farmers and 'half' farmers (i.e. the owners of full taxable units and those which have at some stage in their history been split) are plotted, then the former are seen to lie in an outer band of large courtyard farmsteads, while the buildings of the interior, particularly those on the small 'island' compartments surrounded by lanes, are the properties of the smallholders, who clearly represent later arrivals to the scene. During the nineteenth century this rather large village supported a population of thirty-three full farmers, fifteen half farmers and fifty-six smallholders; by 1861 the village, in addition to farmers, contained a baker, cooper, nautical pilot, glazier, shopkeeper, limeburner, painter, maltster, draper, miller, ropemaker, saddler, ship's carpenter, ship's master, shoemaker, tailor, smith, and a carpenter and timber merchant. The latter was doubtless involved in the construction of many of the fine red and white half-timbered courtyard farmsteads, typical of this part of Scandinavia. The way these are preserved within the fabric of the settlement of 1978 is seen by a comparison of the two maps. In this case historic form and historic fabric are subject to deliberate preservation because their qualities, originally quite work-a-day, are now seen to be both visually attractive and historically interesting. It is inevitable that maps can only create a two-dimensional picture and what has already been said in Chapter 4 concerning vernacular architecture is important here, for this represents a third visible dimension of a village plan. Indeed in Nordby the splendid half-timbered farmsteads are part of the settlement's quality.

The processes changing plans: internal social and economic links

The examples discussed have laid emphasis upon the physical characteristics of village and hamlet plans, but the two detailed studies presented above show that such settlements are inhabited by different categories of people and that their ultimate form may in some way reflect this. Furthermore, as the case of Kiebingen showed, these relationships need not be stable, for replacement populations may be intruded. A most vivid description of what is involved in the presence of internal social and economic links is provided by a description from the work of Italo Calvino – all that needs to be done is to substitute the word 'village' for Calvino's 'city':

> In Ersilia, to establish the relationships that sustain the city's life, the inhabitants stretch strings from the corners of the houses, white black or grey or black and white according to whether they mark a relationship of blood, or trade, authority, agency. When the strings become so numerous that you can no longer pass among them, the inhabitants leave: the houses are dismantled; only the strings remain.
>
> From a mountainside, camping with their household goods, Ersilia's refugees look at the labyrinth of taut strings and poles that rise on the plain. That is the city of Ersilia, and they are nothing.
>
> They rebuild Ersilia elsewhere. They weave a similar pattern of strings which they would like to be more complex and at the same time more regular than the other. Then they abandon it and take themselves and their houses still further away.
>
> Thus, when travelling in the territory of Ersilia, you come upon the ruins of abandoned cities without the walls which do not last, without the bones of the dead which the wind rolls away: spiderwebs of intricate relationships seeking a form.
>
> (Calvino 1979: 61–2)

This is a curious, powerfully imaginative image – 'relationships seeking a form' – an image which touches the intricate networks of linkages found within any settlement, and if instead of a move to another site we envisage a move in time – the passage of a settlement from one time period to another – then the inevitable transfer of important linkages and networks from one period to the next can be glimpsed. Calvino's observation reaches deep into the fabric of a living settlement, beyond walls and mortar, roof or hearth, to involve a community of people, with networks of sustained family and other relationships (Howell in Goody *et al.* 1978: 112–55). The passage also provides a glimpse of the intellectual baggage that emigrants carry with them, so that common English place-names and village forms were translated to North America, German names and farmhouse styles to Australia, Welsh names to Argentina.

The small hamlet of Buble (Figure 5.7 upper) is drawn from a remarkably detailed American study of Ifugao farmers in the hill country of Luzon, in the Philippines (Conklin 1980). Wherever slope gradient and water supply allow, these people construct 'pond fields', terraces with raised, or bunded, edges for the cultivation of rice. These fields, often with many layers of infill and supported by terraced walls, are very valuable properties. They are associated with tiny hamlets, each consisting of one or more minuscule levelled terraces for dwellings or granaries, placed near to the pond fields but on land less suitable for cultivation. These culturally created 'platforms' form the focuses for intense social, ritual and economic activity. The illustration shows, first (A), the layout of the hamlet, comprising a series of house platforms, the rising roman numbers indicating a rise of a metre or so in the terrain. The scale of this drawing is much enlarged relative to the other plans illustrated in this study. Second (B), the buildings of the ten households are framed, with a continuous line showing the kinship groupings between houses; finally, the family composition of each household is shown, using a much simplified genealogical tree spreading across several generations. There are two essential points here: first, even within this apparently simple ten-farmstead hamlet there are vast invisible linkages which to be defined and recorded require the techniques of the anthropologist. Second, the inhabitants themselves are aware of these 'strings' which were an essential part of the fabric of their lifestyle.

Where evidence survives from early European settlements, it shows similar complexities. A classic study of a French village and its inhabitants between 1294 and 1324 is preserved among the records of the Inquisition – for the inhabitants belonged to a heretical sect, the Albigensian Cathars. Montailloux, in south-western France, near the Spanish border, in 1300 comprised about 200–50 inhabitants, divided up into twenty-three hearths or households, clustered on a hilltop below a castle, the houses set one above the other with small gardens, the lowest built sufficiently close together to form a defensive line. The village comprised a set of basic social cells, each consisting of a peasant family, embodied and co-resident in a single house. This, multiplied a dozen times, constituted the village. This was the framework to which individuals were bonded and within which their lives were lived. From each house, as Calvino reminds us, invisible strings stretched to other houses, marking relationships 'of blood, or trade, authority, agency'. In Montailloux as in Luzon the house, *ostal* or *domus*, was the unifying concept of social, family, economic and cultural life.

The very name of Altheim – 'old village' – suggests ancient roots, but the plan (Figure 5.7, lower), involving two major street axes set at right angles, plus an irregular grid of lanes, together resulting in a highly irregular agglomeration, a *Haufendorf*, fails to reveal the true nature of the settlement. The map, however, identifies the status of individual buildings in about 1850. At that stage, whatever the situation had been in earlier centuries, there were only a small number of working farms, forming five small clusters. It is possible that in this we are seeing echoes of the origins of the village, from a series of scattered cells or a linked farmstead cluster, but the possibility that there were once more and that numbers have been reduced as a result of war and reorganisation cannot be excluded. The vast mass of buildings are the dwellings of rather poor smallholders,

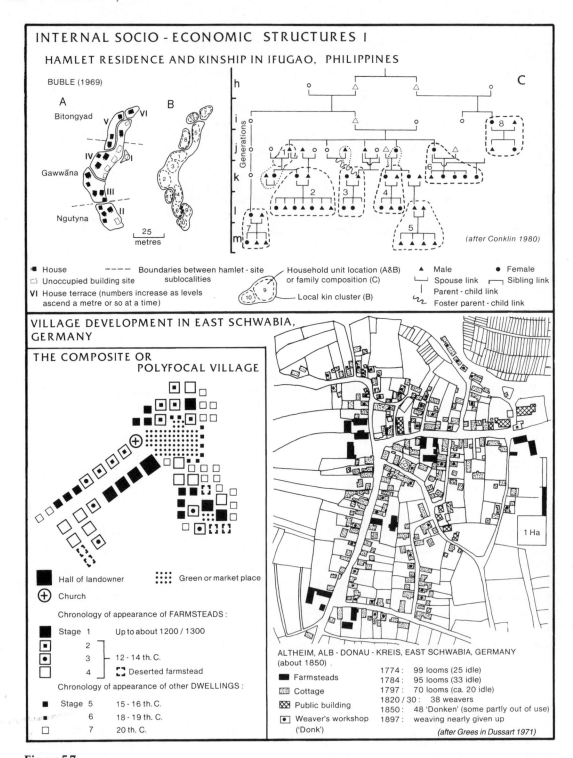

INTERNAL SOCIO - ECONOMIC STRUCTURES I

HAMLET RESIDENCE AND KINSHIP IN IFUGAO, PHILIPPINES

BUBLE (1969)

A

B

Bitongyad

Gawwāna

Ngutyna

25 metres

h
i
j
k
l
m

Generations

C

(after Conklin 1980)

House — Boundaries between hamlet - site sublocalities

Unoccupied building site

VI House terrace (numbers increase as levels ascend a metre or so at a time)

Household unit location (A&B) or family composition (C)

Local kin cluster (B)

Male ▲   Female ●

Spouse link   Sibling link

Parent - child link

Foster parent - child link

VILLAGE DEVELOPMENT IN EAST SCHWABIA, GERMANY

THE COMPOSITE OR POLYFOCAL VILLAGE

1 Ha

■ Hall of landowner   ⋮⋮⋮ Green or market place

⊕ Church

Chronology of appearance of FARMSTEADS :

Stage 1   Up to about 1200 / 1300

2
3   12 - 14 th. C.
4   ⸬ Deserted farmstead

Chronology of appearance of other DWELLINGS :

Stage 5   15 - 16 th. C.

6   18 - 19 th. C.

7   20 th. C.

ALTHEIM, ALB - DONAU - KREIS, EAST SCHWABIA, GERMANY (about 1850) .

■ Farmsteads

Cottage

Public building

Weaver's workshop ('Donk')

1774 :   99 looms (25 idle)
1784 :   95 looms (33 idle)
1797 :   70 looms (ca. 20 idle)
1820 / 30 :   38 weavers
1850 :   48 'Donken' (some partly out of use)
1897 :   weaving nearly given up

(after Grees in Dussart 1971)

Figure 5.7

cottagers (*Seldner*), with little more land than a kitchen garden. Throughout the later eighteenth and much of the nineteenth centuries these people augmented their living by the domestic weaving of both flax and, later, cotton. They tended to concentrate in settlements having a modest centrality, i.e. market villages or little towns, where partible inheritance created a fragmented pattern of tenure and landholding. Such rural industries, primary extraction, the growing of specialist crops such as flax, and the processing of natural raw materials, wood, clay, wool and flax, remained a normal element in many European rural villages long after the main elements of the Industrial Revolution were in place. It was quickening industrial development, the rise of industrial towns and zones and the development of a better rural transport infrastructure which finally brought these supporting activities to an end, leading to decay and shrinkage in many settlements. The *Seldner* of the Ulm region of southern Germany, often working for property-owning master weavers who may themselves have possessed farms, were part of an economic system historically termed 'the dual economy' variously involving farmer-weavers, farmer-lead miners, farmer-quarrymen and even farmer-coal miners (Grees in Dussart 1971: 179–203; Grees in Nitz 1987: 281–94). Of course, the task of earning a living in diverse ways was often seasonal or cyclic, and involved the use of the labour of varied members of the family unit. The ultimate roots of this arrangement are undoubtedly medieval and probably even older. The original farmsteads would have possessed kinship links, and the *Seldner* may have originated as lower-status relatives who were allowed smallholdings in return for their labour.

Throughout Europe in the medieval period a characteristic internal social division within nucleated settlements was between lord and tenants: between the lord of the manor, who held the village from a greater lord or ruler, and his tenants, farmers who were granted holdings for a rent, rendered in the form of work, supplies, labour and, as a money economy developed, cash. The feudal system was a means of supporting a military and ecclesiastical aristocracy by the labour of servile or semi-servile peasant tenants, and for this purpose kings and great barons granted villages and their lands to lesser knights, mounted warriors who formed the backbone of feudal armies. Tenants were recruited in a variety of ways, in return for stock and seed, from emancipated slaves, from less well-endowed kin, and not least as colonists, willing to change their place of residence in the hope of bettering their economic and social circumstances. It is possible that if we could strip away all of the subsequent accretions two fundamental types of nucleation would be apparent underlying the villages and hamlets which are now such a familiar part of the European scene: those deriving from kinship and those deriving from lordship. In practice, however, this must be too stark a division, and both elements must have been present in many settlements. Nevertheless, the contrast is useful, because it is likely that there is also a broad tendency for the regular plans to be linked with lordship, with irregular layouts associated with less formalised kinship accretions.

The formalisation of structural differences deriving from ultimate origin, be these of religion, social class and trade or craft, is also a characteristic feature of the Indian subcontinent. The village of 'Pipersod' (Figure 5.8, in fact a pseudonym) is a rather large irregular agglomeration, sub-divided between several castes, mainly *Brahman*, a high caste, and *Kirar*, cultivators, but also includes Hindu artisan and service castes and a community of Muslims. In detail there is some intermixing of dwellings but, nevertheless, there is a clear tendency for particular social groups to predominate in particular

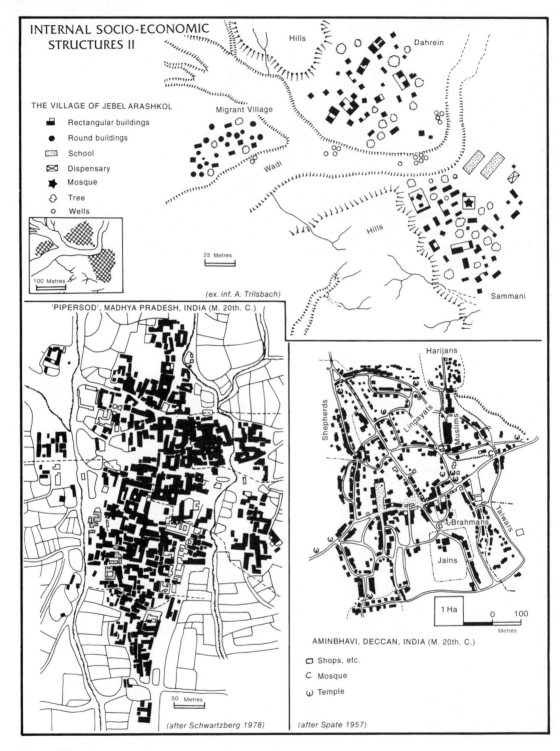

INTERNAL SOCIO-ECONOMIC
STRUCTURES II

THE VILLAGE OF JEBEL ARASHKOL

- 🏠 Rectangular buildings
- ● Round buildings
- School
- ⊠ Dispensary
- ★ Mosque
- 🌳 Tree
- ○ Wells

100 Metres

25 Metres

Dahrein

Hills

Migrant Village

Wadi

Hills

Sammani

(ex. inf. A. Trilsbach)

'PIPERSOD', MADHYA PRADESH, INDIA (M. 20th. C.)

50 Metres

(after Schwartzberg 1978)

AMINBHAVI, DECCAN, INDIA (M. 20th. C.)

Harijans

Shepherds

Lingayats

Muslims

Talwars

Brahmans

Jains

1 Ha   0   100
Metres

- ⌑ Shops, etc.
- C Mosque
- ω Temple

(after Spate 1957)

Figure 5.8

parts of the village. This is seen more clearly in the case of the irregular agglomeration of Aminbhavi (Figure 5.8), although in this example the source plan does not show property boundaries. Finally (Figure 5.8, upper) the village of Jebel Arashkol lies in the Sudan, 7 miles west of the White Nile. The figure is an approximately scaled sketch map, derived from fieldwork by Trilsbach, air photographs and ground-level pictures (Roberts 1987a: 68–71). No less than three separate nuclei are present, and in this there is a glimpse of the way in which large complex nucleations developed through accretive growth to several smaller, socially or economically discrete nuclei. In Jebel Arashkol two of the settlements have specific names while the third is simply a 'migrant village'. This distinctive morphology reflects both the physical environment and socio-economic factors. With a rainfall of less than 250mm the region is arid, but the amount can vary from 5 to 500 mm and during the last decades drought has been an increasing problem. In the 1920s two brothers, members of a distinctive Muslim sect, moved, as the result of a family quarrel, out from a nearby town; the younger settled on the north side of the wadi or dry valley, where well water was available, and Dahrein is derived from his name. The elder brother settled on the south side, about 50 m to the south, and the village name Sammani derives from the sect to which they both belonged. In this settlement a mosque was established, which eventually became both a seat of Islamic learning and a place of local pilgrimage. The twin villages thrived, supported by irrigated agriculture, some rainfall agriculture and associated pastoralism, and eventually housed about 1,500 people. The two primary clusters comprise square mud-brick buildings and small enclosures. In this region round dwellings tend to be the first to appear, built of straw with grass thatch, but they have a short life, and as wealth increases time and labour are invested, on a piecemeal basis, in the construction of more permanent mud-brick dwellings with verandas and courtyards.

By the 1980s marked changes had taken place: the population had reached about 2,700 and a third nucleus had appeared, containing refugee members of the same sect, who preferred not to settle on the edge of the older nuclei. Drought has devastated the farmlands; indeed these are now disappearing beneath sand dunes, and the economy has collapsed. In this almost treeless area the absence of fuel means that the inhabitants cannot even make bread but must subsist upon a form of sorghum gruel, unless fresh loaves are brought from the small town, an hour's drive away. The population is now comprised of the very old, some women and the very young, supported by remittances from menfolk working elsewhere.

Rural settlement dynamics

To understand the evolution of northern European villages it is essential to grasp that many originated as a set of equally sized planned farms, each a standard taxable or fiscal unit. In the course of time these became divided into half, a third, or other proportional or unequal shares. This division could even extend to the sites of the farmsteads. Smallholders or cottagers often represent relative late-comers to the structure of such a settlement, arriving when its territory had already been shared out among the well-established farmers, so they were forced to occupy either small portions of former commons or the deserted sites of former full-status farmsteads. In areas of impartible inheritance the smallholdings could represent the limited footholds of younger children, but in areas of divided by partible inheritance, extreme sub-division could be linked to demographic success within some families, i.e. the presence of many children.

The model in Figure 5.7 (lower, left) generalises this argument. It suggests that large, rather irregular complex plans may be the end-product of a continuous development since the twelfth and thirteenth centuries, with small settlements, the linked farmstead clusters and linked hamlet clusters of earlier centuries, gradually accreting dwellings and changing function to generate the large twentieth-century villages or even market towns. The term *composite* (or *polyfocal*) may be applied to these. With modifications and qualifications this is a picture valid for much of Europe, perhaps having wider implications (cf. Baila, Figure 5.2). The varied parts, plan elements, and basic plan types of which this model is composed, may represent more than historic growth phases; they can represent social contrasts, landownership contrasts, planned or unplanned accretions, growth phases or tenurial distinctions. Moreover, it is possible that forces more powerful than accretive growth act to generate nucleations, the factor summarised above as 'communality of enforcement'.

Villagization or congregación

While the social and economic forces found within a community are clearly important in generating nucleations, there may also be external pressures, with political power creating settlement nucleations for many reasons – to settle colonists, to control, to proselytise, or for ideological purposes. There are documented cases of villages and towns being established throughout eastern Europe during the thirteenth century as a result of the 'drive to the east' by the Teutonic knights and warring bishops, on whose lands German and Flemish migrant farmers were encouraged to settle (Simms in Smyth and Whelan 1988: 22–40). Regular planned villages were established, often using a land agent or locator as an intermediary. In Scandinavia there are grounds for suggesting that the regular planned village layouts of parts of Denmark and Sweden resulted from the demands of royal taxation. In England, and in measure Wales and Scotland, the growth of villages, both geometrically planned and of more haphazard form, are linked with the administrations of the great estates of the medieval period, whose lords needed surplus grain, wheat, rye and barley, to provide bread and beer for their great households, and oats to feed their numerous horses. These, along with sheep, were essentially the products of stiff loam and light loam regions, the limestone escarpments and clay vales of southern England, where population levels in 1086 were still relatively low, i.e. in what became the great village belt (Figure 3.2). The villages and farms were turned into a great swathe of grain-producing enterprises, the productive capacities of each settlement, each parish and each township being harnessed to this end. Thus great estates represent a fundamental context where many aspects of settlement were generated. Indeed they may be the progenitors of true villages throughout much of Europe.

Further to this, communality of enforcement was a process found in other historic contexts. Following their conquest of the Inca civilisation of the high Andes the Spaniards congregated the peasantry into villages, a means of control which had disastrous economic results, for it led to the abandonment of the ancient irrigation systems and terraced fields. To this process the useful Spanish term *congregación* may be applied. Figure 5.9 incorporates material from two more recent cases, Malaya and Tanzania. Before and during the Second World War difficulties arose over many

thousands of Chinese squatters, originally brought in as labour for the rubber plantations. Forbidden to own land, they were forced into illegal occupation, squatting (Sandhu
in Dale and Kinloch 1963/4: 157–83). In 1948 there were at least 300,000 of these people,
increasing in numbers and prosperity, but inevitably emerging as both a political force
and a security risk. During the war the Malayan Communist Party had, with the
connivance of the British, retreated into the jungle to build up an anti-Japanese resistance
movement and after the war the squatters provided a fertile ground for their activities.
Scattered in the jungle, on the fringes of the rubber estates, the squatters, working by day
as rubber tappers, market gardeners or miners, provided ears, eyes and a smokescreen
for the Communists and at night became active guerrillas. As is normally the case,
coercion and conflict intermingled with legitimate objectives. At first the government
tried the repressive policy of uprooting the squatters, detaining and then deporting
them, but it was soon realised that a more constructive plan was required. Between 1949
and 1950 varied local schemes for resettling these groups were attempted, but after 1950
sustained policies of relocation and regroupment were followed. Relocation was a more
general policy of bringing squatters together at a fortified site, sometimes new,
sometimes adjacent to an existing village, while regroupment involved drawing
dispersed mine and estate labourers, their families and their dwellings, to a fortified
point on or close to the estate of the employer. Nearly 573,000 people were moved,
300,000 squatters and 273,000 legitimate occupiers, of whom 86 per cent were Chinese, 9
per cent were Malay and 4 per cent were Indian. Some government funds were made
available to ease the transition, but there were inevitable hardships. The standard 'new
village' is seen in Figure 5.9, although only one-third were in fact wholly new. At
foundation many resembled closely packed shanty towns, and there was serious
disruption to agriculture. Nevertheless, some 600 'new' settlements were created, and
this was considered to be the decisive move in the successful operations against the
Communists. As in Malaya – now Malaysia – for census purposes a concentration of 1,000
or more inhabitants was in the 1950s considered urban, villagization generated potential
for the growth of rural market towns. A similar policy was pursued by the British during
the Mau Mau emergency in Kenya.

In practice villagization is a complex process, and in Tanzania between 1973 and 1976
much of the rural population was consolidated into settlements of roughly 250–600 households, although some moves had been made in the previous three years. This villagization
was the result of Tanzania's ideology of 'ujamaa' – literally 'familyhood' – derived from the
ruling Party of the Revolution and articulated by Julius Nyerere. Figure 5.9 (lower) trans-
lates the matrix of ideas into a diagram, structuring the ideas present in the ideological and
practical roots. The impact upon local communities were many: some of the changes were
beneficial, but others, particularly for the women, were detrimental. As was the case in Fiji
(see Figure 6.3 below), relocation relative to roads expanded the potential for market production, opened the possibilities for casual work and stimulated small industries, pottery,
fish-drying, poultry production, handicrafts, tailoring, charcoal production and, not least,
brewing, although increased access also allowed the import of cheaper manufactured
goods. In the short term then, the results are variable, but undoubtedly in the longer term
the nodes created may become part of a new and stable system.

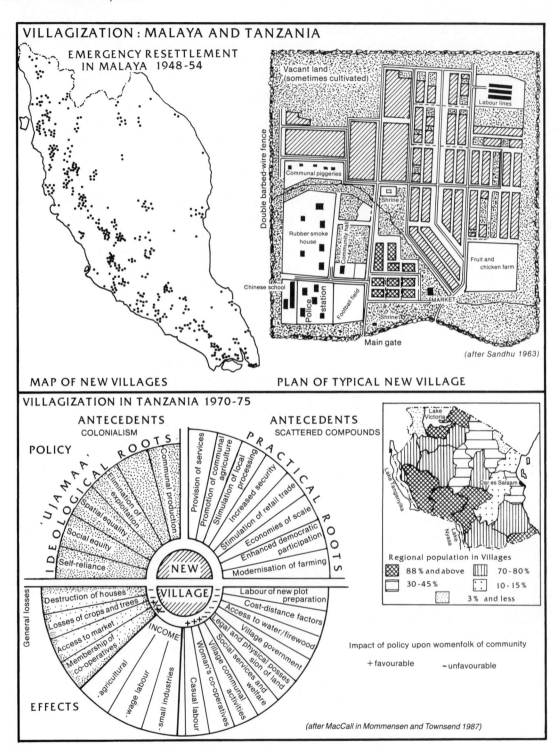

VILLAGIZATION: MALAYA AND TANZANIA

EMERGENCY RESETTLEMENT IN MALAYA 1948-54

MAP OF NEW VILLAGES

PLAN OF TYPICAL NEW VILLAGE

Vacant land (sometimes cultivated)

Labour lines

Double barbed-wire fence

Communal piggeries

Shrine

Rubber smoke house

Community hall

Fruit and chicken farm

Chinese school

Police station

Football field

MARKET

Shrine

Main gate

(after Sandhu 1963)

VILLAGIZATION IN TANZANIA 1970-75

ANTECEDENTS
COLONIALISM

ANTECEDENTS
SCATTERED COMPOUNDS

POLICY

'UJAMAA' IDEOLOGICAL ROOTS

PRACTICAL ROOTS

Elimination of exploitation
Communal production
Provision of services
Promotion of communal agriculture
Stimulation of local processing
Increased security
Spatial equality
Stimulation of retail trade
Economies of scale
Social equity
Enhanced democratic participation
Self-reliance
Modernisation of farming

NEW VILLAGE

Labour of new plot preparation
Cost-distance factors
Access to water/firewood
Village government
Legal and physical possession of land
Social services and welfare
Village communal activities
Woman's co-operatives
Casual labour

General losses
Destruction of houses
Losses of crops and trees
Access to market
Membership of co-operatives
INCOME
- agricultural
- wage labour
- small industries

EFFECTS

Regional population in Villages
88% and above
70-80%
30-45%
10-15%
3% and less

Lake Victoria
Lake Tanganyika
Lake Nyasa
Dar es Salaam

Impact of policy upon womenfolk of community

+ favourable   – unfavourable

(after MacCall in Mommensen and Townsend 1987)

Figure 5.9

Nucleation and dispersion: trajectories of change

A number of the individual cases discussed have emphasised that villages often appear to develop from smaller, simpler arrangements, either hamlets or single farmsteads, and the hierarchy of settlement types at the lower levels is more diverse in character than the terms single farmstead, hamlet, village and town imply. Irrespective of their detailed morphology, which may be regular, irregular or lie at a point on the catena between these extremes, Table 5.1 suggests the diversity present at the lower levels of the hierarchy, between single farmstead and town, and suggests some possible development trajectories, likely directions of change when one settlement type is successful, and by adding new dwellings achieves a change in form and sometimes function.

The right-hand column of Table 5.1 indicates likely directions of change. For instance, a hamlet (6) may grow and be transformed into either a group of interlinked hamlets (7) or a single-plan type village (8), perhaps an irregular agglomeration: this in turn may, in the right economic circumstances, develop into a larger composite plan (10); if this acquires a market (11) it may eventually achieve a secure position in the hierarchy as a market town. This rather skeletal model serves three purposes. First, it does express an idea with clarity, while, second, it is a generalisation applicable in many contexts at many scales, probably having universal validity; third, it gives a brief glimpse of the variety of changes that are possible within a single system of settlement throughout time. Furthermore, this change need not be unidirectional, so that individual settlements, when subjected to delay and depopulation, may also pass downwards in the table.

These ideas are built into the more graphic model in Figure 5.10 which allows a very

Table 5.1 Types of settlement and possible developmental trajectories

Settlement types		Possible developmental trajectories	
Level One: Non-nucleated settlement elements			
1	Single farmsteads	1 ==>	2/3/4/5/6
2	Joint farmsteads	2 ==>	5/6
3	Kinship clusters	3 ==>	5/6
4	Linked farmstead clusters	4 ==>	7
5	Dependency clusters (high-status farmstead ('magnate' farmstead, hall or manor hour) with dependent dwellings for farm servants etc.)	5 ==>	8
Level Two: Nucleated settlement plan types			
6	Hamlet	6 ==>	7/8
7	Linked hamlet clusters	7 ==>	8/9/10
8	Single-plan type village	8 ==>	10/11/12
9	Village with closely dependent but separate hamlets	9 ==>	10
Level Three: Complex nucleated plan types			
10	Composite or polyfocal village	10 ==>	11/12
11	Market village/trading village/quasi-urban village	11 ==>	12
12	Market town		

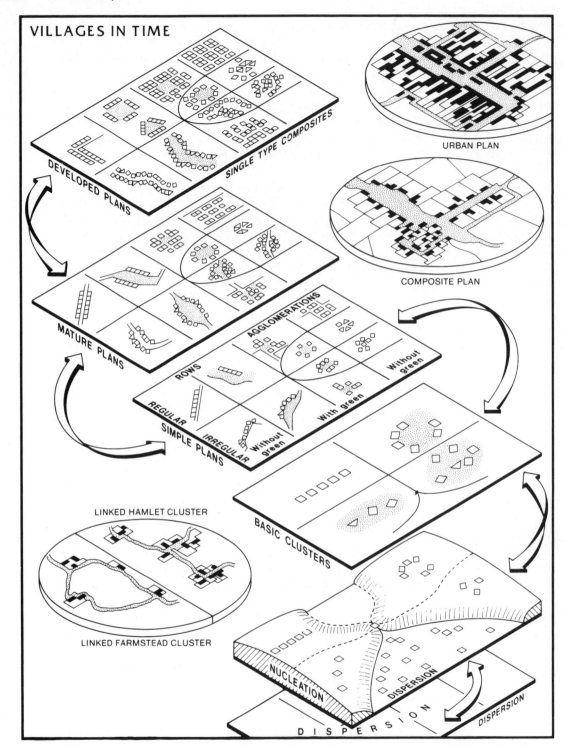

VILLAGES IN TIME

DEVELOPED PLANS

SINGLE TYPE COMPOSITES

URBAN PLAN

COMPOSITE PLAN

MATURE PLANS

AGGLOMERATIONS

ROWS

REGULAR

IRREGULAR

SIMPLE PLANS

Without green

With green

Without green

LINKED HAMLET CLUSTER

LINKED FARMSTEAD CLUSTER

BASIC CLUSTERS

NUCLEATION

DISPERSION

DISPERSION

DISPERSION

Figure 5.10

broad view of change in rural settlement to be depicted as a matrix: it is in fact a version of the grid, Figures 5.1 and 5.2, and shows a series of stylised hamlet and village plans on 'plates'. The one labelled 'mature plans' corresponds to that shown more fully in Figure 5.1, but extends the idea by recognising that in the real world simple versions of the mature plans can also be found, and as the cases and the models already discussed clearly show, simple plans can develop into mature plans (Figures 5.5 and 5.6). Mature plans, either larger, often single-type composites, with an intensification of plot use, are characteristic of the village/small town threshold. Thus an examination of a single cell of the grid, passing from simple, to mature and thence to developed, generalises the through-time trajectory of a single settlement. The simplest level of small nucleations – and the terms used in this discussion avoid the words 'village' or 'hamlet' – have been designated 'basic clusters', small groupings of a few farmsteads or homesteads, very simple, while the final two plates touch the nucleation/dispersion threshold.

In one sense this diagram, based upon the examination of many thousands of British examples, is wholly theoretical. Nevertheless, it defines the possible time trajectories of all types of nucleation and deals with all forms of settlement from the single farmstead to a small town, and creates a framework within which to envisage processes of change within settlement forms. It identifies two dimensions of change: first, major changes, involving shifts from one level in the hierarchy to another, and second, minor changes, which although limited in scale, affect individual nucleations differently, but which, cumulatively, can result in radical developments.

Figure 5.11 is less theoretical, and was devised to show what is currently taking place within the villages and hamlets of north-west Europe. It suggests that future developments may follow four separate pathways. These towards contraction and eventually total destruction were plotted with the policies being pursued in Romania during the later 1980s by the Communist dictator Ceauşescu, a project now wholly overtaken by events. The more recent devastation of the Kurdish settlements in Iraq and villages in the former state of Yugoslavia is a reminder, however, that, in the right political circumstances, total settlement destruction is never wholly impossible. Nevertheless, if we take a wider view of settlement development, and ignore the extremes such as total destruction, then three broad processes of change are taking place within inherited rural settlements, summarised by the words 'stability', 'expansion' and 'contraction'. These processes are generated by various forces, some being unplanned, while others are part of policies imposed by national or local government. They are summarised in the grid at the top left of the model. At the bottom left is a reminder of the historic factors creating social and economic contrasts (Mills 1980: 117). Of course, this model is concerned with nucleation, where the social and economic differences are the more sharply apparent, but in areas of dispersion and hamlet settlement the forces generating the contrasts are less focused, more diffuse, so that the problem is one of identifying and explaining small-scale regional variations rather than differences between individual settlements.

Stability – a condition of only limited change such as additions to existing buildings and small-scale plan changes such as road improvement – may result from neglect, perhaps the result of a locality's marginal geographical location. In contexts where there are few pressures to generate expansion, i.e. an absence of employment, the survival of an ageing local population and perhaps limited tourism are sufficient to sustain the general character of the inherited forms. Such factors preserve many small nuclei in the remoter upland regions of Britain. More rarely, and cases are not unknown, the policy

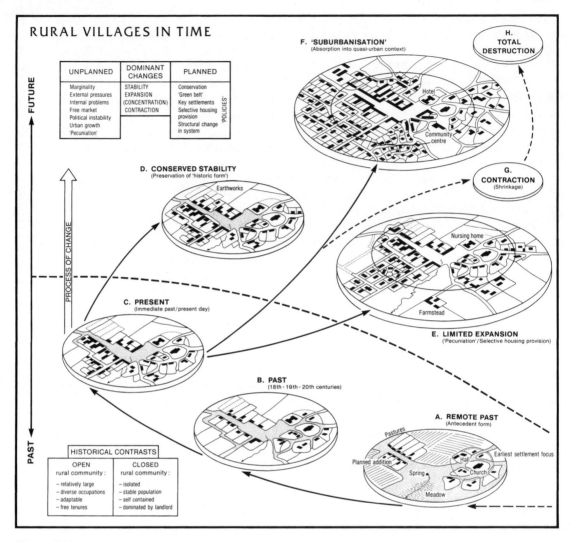

Figure 5.11

of a landowner is such that stability or indeed actual neglect may be imposed. On the other hand, the combination of an attractive inherited plan, buildings which are survivals of high-quality traditional styles, and an advantageous location relative to large towns providing employment may be powerful enough to have a settlement declared a conservation area, with restrictions and controls imposed upon development proposals. It must be said that economic wealth and concentrated local political power are also necessary ingredients for the sustained and successful implementation of such a policy. The presence of a great house, with a powerful and resident landowner, the village dwellings held by tenants, is also a powerful preservative, a continuation of the 'closed' villages of the eighteenth and nineteenth centuries thoroughly documented by Mills (1980). In contrast, much freehold, many owners, and a long tradition as an 'open' village will tend to encourage modern speculative development.

Limited expansion is a most common fate of former rural nucleations. This normally occurred during the nineteenth and twentieth centuries, as many large rural houses of that period attest, the dwellings of entrepreneurs, merchants, bankers and other professional men, often located to take advantage of the coach or pony and trap – and eventually the motor car – to give linkage to the local railway station, where designated carriages ensured social separation during the journey to town (Figure 5.5, upper). In more recent times, particularly in the 1980s, the increasingly numerous wealthier elements of the population of England and Wales have continued to desert the cities and industrial conurbations to live in market towns and wholly rural areas. At this time the fastest-growing rural populations were to be found in areas such as East Anglia and Cambridgeshire, south-central England (Northamptonshire, Oxfordshire, Buckingham-shire, Wiltshire, Somerset and Dorset), Kent, East Sussex and Essex, the west Midlands (Shropshire, Powys, and Hereford and Worcester), and to a lesser degree Lincolnshire, North Yorkshire, Cumbria and Northumberland (OPCS 1991). All of these rural regions are now experiencing, to a greater or lesser degree, what this author terms 'pecuniation', i.e the investment of capital in the restoration of the inherited older rural building stock and the construction of new elements intruded into the fabric of ancient plans by developers, creating idealised if distorted images of rurality – really the urban estate projected into a field. Such development is not wholly different in kind from the fields of houses created as mining villages during the nineteenth century. As a term, 'pecuniation' is to be preferred to 'gentrification', for the latter should be reserved for the process in towns by which older areas are 'revitalised' by developers attempting to introduce new and expensive residential zones. Of course, such changes are particularly noticeable when they are concentrated in the villages and hamlets, but the process has also affected dispersed dwellings, and is seen in the widespread improvement of farmhouses, cottages, and barns and stables (see Chapter 2) which are no longer needed for agriculture, and are converted into either permanent dwellings for commuters or holiday cottages.

Finally, those rural villages lying within the hinterland of either long-established or new towns will inevitably experience a greater degree of change than is occasioned by pecuniation (OPCS 1991). The whole fabric of the historic settlement will be changed, metamorphosed into a portion of a growing urban area, yet preserving, in its layout and a few buildings, traces of its rural origins. The urbanised rural village, still preserving its proud identity, is a common phenomenon throughout Europe.

Conclusions

Settlement forms are valuable as historical evidence, but as we now see them they are filtered through centuries of change and adaptation. These do not, of course, appear within a vacuum; they function within distinctive contexts, patterns, of which the forms are the constituent parts. It is to these broader frameworks that discussion must now turn.

CHAPTER 6

Settlement patterns

Individual human habitations can be divided into several categories which reflect their degree of permanence, from the wholly ephemeral, of merely a few days' duration, to permanent structures which have had a long existence within human memory (Chapter 2, above). Of course, no buildings last for ever, but standing Roman and Greek structures and the great castles and churches of medieval Europe suggest that if they are well built, maintained and, of course, not knocked down, they can survive for millennia. Nevertheless, the terms 'short-lived' and 'long-lasting' establish two ends of a scale which can be used to describe either the life of a building or a settlement's duration in time, and this idea is fundamental to understanding the character of patterns, i.e. the spatial distribution of the elements of settlement, be these farmsteads, hamlets, villages or market towns, throughout a landscape or region and ultimately the whole earth's surface. The first part of the chapter will develop this idea and define a twofold classification of settlement patterns on the basis of the dominant processes operating within them, give some detailed examples, and conclude with a general model which demonstrates some of the key processes which underlie the evolution of patterns. The second half will examine a series of figures which draw together at the same scale cases taken from three continents and finally interlock the study of forms and patterns using examples drawn from Denmark, Wales and Malaya.

Types of settlement pattern: unstable and quiescent

Figures 6.1, 6.2 and 6.3 are a practical step in the classification of settlement patterns, and distinguish between two fundamental structural categories. The first of these may be termed *unstable*, i.e. those patterns in which the rate of change is such that the overall pattern can be seen to alter substantially within a relatively short time-span, a year or a few decades, while the second may be termed *stable* or *quiescent*, i.e. patterns whose structural arrangements appear to change little within the span of a single human lifetime. Of course, new dwellings, indeed new settlements, may appear, while others may disappear, but an essential stability predominates. However, a word of caution is

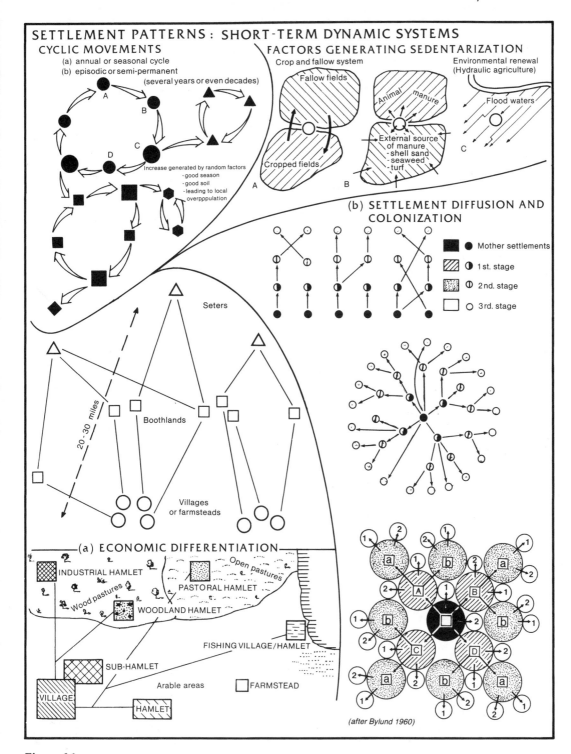

Figure 6.1

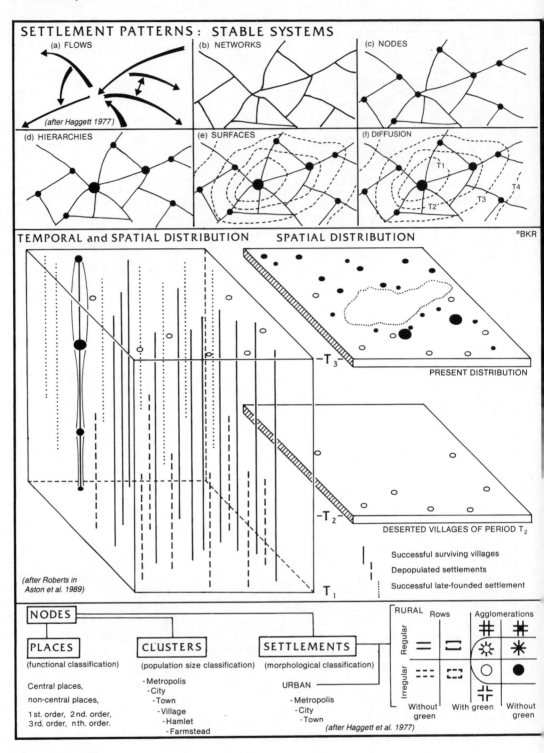

Figure 6.2

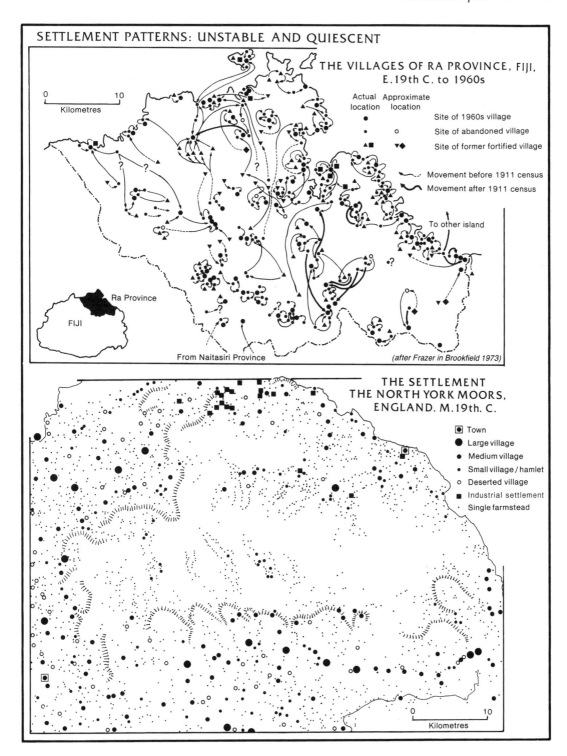

Figure 6.3

needed; we must not, in choosing a term to describe these – quiescent – assume that they will not be subject to drastic change in the future and have not been subjected to equally drastic change in periods long before we can study them. In converse, even within unstable, mobile systems, some elements may remain wholly stable: for example, the same settlement sites may be periodically reoccupied, indeed the population may limit themselves, or be limited, to the same territory, by agreement or due to the presence of powerful hostile neighbours. Within a group's territory a sacred site may be used as a focus for ceremony and ritual – here one can think of sacred sites such as Stonehenge or the great neolithic communal tombs – for centuries rather than decades (Burl 1987; Hedges 1984). Once again, this classification moves beyond mere static description and time is a crucial ingredient of the argument.

Figure 6.1 presents a series of ideas about patterns in which instability is present: the first diagram (Figure 6.1, 'Cyclic movements') shows a situation which may have existed when the first farming communities were breaking into the woodlands of Europe, a landscape up to that date only exploited by hunter–gatherer–fisher communities. Felling and burning the forest trees to create clearings, they planted and raised grain crops in the ash-enriched soils. After a few years yields steadily decreased as soil nutrients declined and there was a strong incentive to move to find and cultivate a new area. There are some grounds for suggesting cyclic movements throughout definite territories, with periodic reoccupations of former sites once the natural vegetation had begun to re-establish itself. However, we must be careful not to see these groups as 'primitive', for recent work has shown that not all neolithic agriculture was at this 'slash and burn' level, and it is quite clear that the farmers quickly developed techniques of stall-feeding stock with fodder derived from the nutritious lopped branches of trees. The stall-manure (a sample has been found in a peat-bog from Switzerland) contained the pupae of stable flies, not those to be expected if it had been voided in the open (Champion *et al.* 1984: 121–51; Evans 1975: 112–57; Renfrew 1987: 120–37). It will be noted in the diagram that now and then settlements – represented by the circles, squares and triangles – sometimes sharply increase in size as a result of random factors such as a 'better' year in terms of yields caused by chance variations in weather or soil. At this point new settlements are hived off. Viewed over a century or so, this type of settlement system is seen to be in constant movement, for the dwellings moved with the fields.

Of course, a settlement system may be unstable for other more complex reasons and Figure 6.3 give an example from northern Fiji. Frazer, an anthropologist, has mapped settlement changes between the middle of the nineteenth century and the 1960s (in Brookfield 1973: 75–96). Fiji became a British colony and this had profound effects upon the traditional society. Before 1874 the inhabitants lived in fortified villages, a situation Frazer describes as 'absolutely essential to the preservation of life', and villages were located upon defensible sites, often a craggy hilltop but occasionally a moated coastal situation. This need for defence led to the abandonment of other earlier settlements. After pacification in 1874 defensible conditions were abandoned for more convenient sites where cultivable land was available for gardens, or close to the sea. Initially these shifts led to the emergence of smaller villages or hamlets, the result of splitting the fortified villages. However, wise choices were not always made, and some settlements were moved from damp and unhealthy sites by order of the provincial medical officer, while others moved spontaneously to be adjacent to newly constructed tracks or were reamalgamated to create larger units. This process of amalgamation and limited movement continued into the twentieth century, so that the eighty-seven villages present in 1901 were further reduced to seventy-eight in 1960, a process largely, according to Frazer, occasioned by improving communications and access to markets. Most new villages were either on or close to roads. It is now likely that an increased awareness of the commercial value of land will prove to be a fixative of the 1960 pattern, so that a relatively quiescent pattern has been achieved. It should be stressed that the word 'quiescent' does not exclude the possibility of future change. What has altered is the degree of stability within the whole pattern, and this case illustrates something of the matrix of social and economic forces at work. It may also provide a glimpse of what is an essentially unstable 'prehistoric' type of settlement system.

In stark contrast, Figure 6.2 models an indisputably quiescent pattern. The top of the box or column (T_3) represents a piece of territory, perhaps 20 by 20 kilometres square, whose most recent settlements are shown on the separate map to the top right, labelled 'Present distribution'. The depth of the column beneath this upper level represents time, perhaps a 1,000 years, in which some settlements are seen to be sustained continuously while others are destroyed and depopulated and yet others are established. The lower map, T_2, records the settlements destroyed by that period, and these also appear on the upper map, at T_3, as open circles. To take the case of England (see Figure 3.2) the figures currently available suggest that during the last 1,000 years or so between one-fifth and one-quarter of all the pre-industrial nucleated villages and hamlets in England have been depopulated. Calculating the additions is more difficult, but the process of industrialisation undoubtedly brought many hundreds of new arrivals. Nevertheless, this time matrix hints at the complexities which underlie all present-day patterns in regions dominated by largely stable quiescent systems. Furthermore, as the single isolated case suggests, an individual settlement can itself undergo many changes, expanding and contracting through the centuries, and the example chosen was once, two or three centuries ago, very much larger than it appears on the modern map.

It is within such mature quiescent systems that the classic features of geographical study, nodal regions, appear in their most fully developed forms. This is illustrated at the top of Figure 6.2 using a model derived from work by Peter Haggett (Haggett *et al.* 1977: 6–10). The six diagrams emphasise that all human societies involve spatial interactions or *flows*, ranging from the exchange of prestige goods and women at the primitive level to the vast tides of goods, people and ideas in the modern world. Such flows follow routes or *networks*, from forest tracks to railways and radio waves from communication satellites. These meet at individual settlements to form *nodes* or meeting points located on the networks. Because these nodes are never uniform in character, some possess more advantages than others, and so a system of articulated *hierarchies* appears, larger settlements in some way dominating smaller ones. The terms applied to these nodes and three ways of classifying their character are placed at the foot of Figure 6.2. The integration of these elements – flows, networks, nodes and hierarchies – may be thought of as fields of influence extending out from the most important settlements and over both groups of smaller settlements and their territories. Variations in these areas allow *surfaces* to be identified, spheres of influence, types of land-use, while the final cell is a reminder that time is always present, for influences, ideas and cultural practices *diffuse*, spreading outwards along definable networks.

Figure 6.3 includes a complete map (at the same scale as Ra Province) of the settlements in and around the North York Moors – a quiescent landscape, including both nucleated and dispersed elements. This, it must be stressed, is only part of a broader pattern which extends across northern England (Figure 3.2). To all intents and purposes this mid-nineteenth-century pattern is identical to that of today: some farmsteads will have been depopulated (but are usually seen as ruins in the landscape), but many of the villages and hamlets have grown, while all the towns will have expanded. Open circles record the presence of deserted villages, either depopulated during the fifteenth century as a result of the conversion of arable land to pasture to carry more stock, or more gradually as a result of the engrossing of holdings into the hands of a few tenants. Whatever the complications, the vast majority of settlements on this map have older roots, for their names appear in 'Domesday Book' in 1086 and many indeed are of Anglo-Saxon or Scandinavian origin. However, this map does raise further questions: the settlement of the central core of the region, high open moorland country with infertile soils developed on Jurassic sandstones, is dominated by dispersed farmsteads. These often form pairs of lines because they tend to be set on the valley sides above the arable lands and meadows, at or just within the head-dyke, the wall or bank and ditch separating the improved and enclosed lands from rough grazings on the higher open land. There is logic in this: the manure from the stalled cattle could more easily be carted downhill for spreading, while stock could easily walk upwards to the rough grazings. Where the River Esk cuts into the uplands, running from east to west in the north-eastern sector of the map, some nucleations appear, but most of these grew up during the nineteenth century after the railway reached the valley.

In general, a tide of villages and hamlets sweep around this great moorland core (Spratt and Harrison 1989: 72–112). To the west, in the Vale of York, a dense pattern features a significant number of depopulations, with one, two or three large farmsteads now working the former village lands. To the south and east nucleations have a complex relationship with the northward-facing escarpment of the Tabular Limestone, with a chain of villages and hamlets along the tail of the dip slope where this meets the flat Vale of Pickering. These occupy a zone of streams, springs, varied soils and gently sloping land which has long been a 'preferred settlement zone', occupied from prehistoric times onwards. The roots of the pattern now visible, however, seem to be medieval, for, as was noted above, the names of most of the nucleations on the nineteenth-century map were already present by 1086. A further clue to the stability of this pattern is provided by the Norman fabric present in village churches, a style generally dating before the closing years of the twelfth century, although at that stage Norman attitudes and building practices, and Norman ideas of village planning and organisation, were being diffused through a countryside moulded by Celtic and Anglo-Saxon communities. Once again the visible pattern is clearly the end-product of centuries of development (Spratt and Harrison 1989). Relative permanence is a result of the development both of a system of farming which allows the fertility of the land to be sustained and of systems of land-use and landownership which vest rights in individuals and preserve written records of these rights.

Returning to Figure 6.1, 'Factors generating sedentarization' suggests that settlement sedentarization can occur in several ways. Certain environments experience natural renewal, classically seen in the great old world riverine regions of Mesopotamia and Egypt, where natural fertility was renewed by flooding. Both are regions whose grain crops nurtured large populations and early civilisations (Van Bath 1963: 7–25; Proudfoot and Uhlig in Buchanan *et al.* 1971: 8–33; 93–125), although in practice these were difficult environments, demanding careful organisation and control, hence a well-developed social structure. Soil fertility can also be sustained by the addition of manure from stalled animals, a technique which was, as noted above (p. 124), already present in temperate Europe by the neolithic period. Manure can be sought from other sources; for example, ash, litter, turf, shell sand or seaweed could be applied to the arable to help sustain fertility, although usually animal manure is a necessary element, the other items being supplements (Proudfoot in Buchanan *et al.* 1971: 8–33). Finally, and throughout northern Europe more usually, arable land can be rested cyclically in a crop–fallow system. The duration of the period of rest can vary from many decades in the case of bush-fallowing, where woodlands are allowed to redevelop, to merely one year, with this fallow kept clear of weeds by means of ploughing, a process which also allows frost and soil organisms to break down the soil minerals and release plant nutrients. One or more of these developments are fundamental to the creation of farming systems which sustain both soil fertility and an acceptable level of crop yields, for if continuous cropping without manuring is practised, then the seed:yield ratio will gradually drop to a level at which it will then be sustained. If fallow or manure is introduced – and this generalisation applies equally well in hydraulic environments where 'natural' fertility is renewed annually – then the seed:yield ratio will rise, i.e. a single unit of seed sown will produce a greater quantity of grain at harvest (Van Bath 1963: 18–23). The importance of grain, nutritious, capable of surviving storage, useful as a commodity for trade or the exercise of political power (it will feed armies), cannot be over-stressed: it is the foundation-stone of civilisation, for the need to measure land, count and record the harvest, store and then reallocate, represented powerful forces generating both numeracy and writing.

Key processes affecting developing patterns

It is a common characteristic of settlement studies that two extremes, such as stability and instability, can often be defined, but this must then be qualified by pointing out that the contrasting characteristics vary through both space and time. Instability can give way to stability or vice versa, at different times and in different regions. In practice any unstable, developing settlement pattern will tend to be moulded by the action of four factors (Figure 6.4): the presence of continuity, the presence of rapid, often cataclysmic change, the impact of colonisation, and the existence of changing economic systems, which over many centuries may be identifiable as a succession of systems associated with economic differentiation and demographic change.

Continuity and cataclysm

The idea of 'continuity' within a settlement system is at once simple and subtle: the word is used to suggest that the settlements making up the pattern, however they may change individually, nevertheless remain upon long-occupied sites, so that these parts of the overall pattern remain stable throughout many centuries. This idea has already been introduced in Figure 6.2 and was used in the discussion of settlement around the North York Moors. Of course, within the overall pattern certain components can disappear and new ones be established, but the essential emphasis is upon stability. Sustained

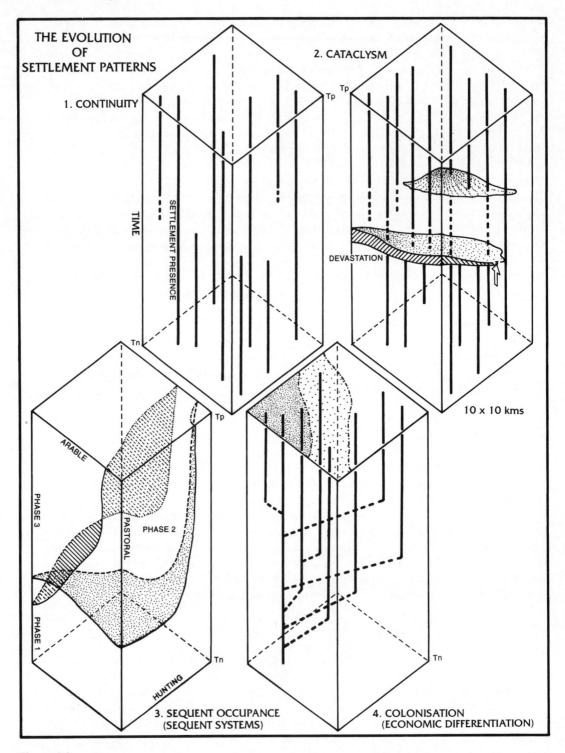

Figure 6.4

continuity of the nucleated components is a characteristic of mature and stable systems, where investment in the infrastructure has been sustained throughout many centuries. Awareness of the likely presence of continuity has an effect on thinking about settlement: it is only within the last forty years that the slight traces of prehistoric, Romano-British and Anglo-Saxon settlement have been recovered from beneath surviving or deserted English villages, traces which must represent only a small proportion of what is in fact present. What this 'continuity' really represents in terms of economy, society and population remains an open question, for it is difficult to prove absolutely continuous occupation using archaeological methods. Nevertheless, the recognition that sites still used and occupied were also part of long vanished settlement systems, the most obvious traces of which may be deserted archaeological sites, has a bearing on the interpretation of archaeological distribution maps and the interpretation of earlier settlement systems.

Most settlement processes tend to be gradual, having an effect over many decades or centuries, but some processes can operate very rapidly, with changes taking place within a few decades, years, or even days or hours. Geomorphological changes, such as the alteration of a river course, can take place within a few hours, but culturally induced cataclysms can also be rapid, for instance the pattern of villages seen in North Yorkshire in Figure 6.3 was imposed over a landscape devastated during the terrible winter of 1069–70 by a Norman army suppressing rebellion. Such devastations have been a recurrent feature of settlement development throughout time, and for the indigenous inhabitants of any country, foreign invasion often induces phases of catastrophic change. This is an experience common to regions which have experienced colonialism, a theme discussed in Chapter 7.

Colonisation

Colonisation involves the diffusion of new settlement or new types of settlement into regions where it was not formerly present. Figure 6.1, 'Settlement diffusion and colonisation', derives from models developed in Swedish Lapland by Bylund (1960), created when he was trying to assess the stages by which land of varying qualities was occupied. While he stresses that these models are of 'purely theoretical concepts', they have attracted much interest. Two examples of his theoretical stages are shown, two of which are concerned with outward expansion from a single settlement and with expansion from a line of mother settlements, as along a coastline. The final model – bottom right – so often not republished by commentators, leaps ahead conceptually, and shows a refined version which Bylund considered to be a close approximation to reality. The availability of good land is not the only problem; it has to be present in sufficient amounts to support farmers at a specific level of technology, from shifting agriculture to fertiliser-sustained rotations. In Bylund's specific example, in an eighteenth- and nineteenth-century context, the introduction of irrigation to improve hay production and of potatoes as a food crop made it possible to make a living from smaller and smaller holdings. It is these smaller units which form the latest generation, both consolidating the settled area by infill within areas already settled, and extending the margins of the settled area. Nevertheless, it should be emphasised that this stimulating model is derived from a region wholly at the margins of permanent settlement in Europe. Furthermore, the argument begins with initial nodes established before 1775, and documents a colonising process sustained throughout the later eighteenth and nineteenth centuries.

Sequent systems: economic differentiation

The model in Figure 6.4 also contains the idea, bottom right, that as economic systems develop, so settlement characteristics adjust. The projection of the three simple systems to the upper surface of the time–space column is a reminder that all territories need not be uniformly affected by these developments, and that for instance uplands may preserve more archaic types of economy than developed lowland: the western uplands of county Durham were used as hunting preserves by the medieval bishops of Durham, preserving land usages which probably had roots in prehistory, but echoes of this former land-use are still to be found in the syndicated shoots which blast away at the humble grouse every August. In practice, colonisation was also associated with a process of economic differentiation, and Figure 6.1, 'Economic differentiation', is a model based upon work done in north-central Sweden.

To grasp the significance of the upper, more skeletal diagram, it has to be appreciated that most of Sweden is a difficult landscape. The limited areas of cultivable soils must be extended by back-breaking land clearance, removing the vast quantities of erratic blocks which mask even potentially cultivable land, and this itself is often pinched in between bare or, almost bare, rock, boulder-strewn woodlands, marsh and lake. Land improvement has been a slow, accumulative process, so that many of the present-day 'good' soil areas are in origin often prehistoric fields. On the more extensive areas of better land loose groupings of farmsteads emerged, from Viking times and earlier – often termed 'villages' – but corresponding to the 'linked farmstead clusters' noted in Chapter 5. Sustained by limited arable, garden plots, hay meadows and pasture areas, these settlements at first utilised only the lands in their immediate vicinity, but population increases necessitated the reclamation and use of lands further afield, at first for summer grazing purposes, with cattle being moved out in the spring and back to the home-farm in the autumn. Between these 'seters', which eventually could often be rather distant, and the home settlements appears a category termed 'boothlands', areas where some arable, with meadowland and woodland resources, could be exploited, but which were worked from the home farmsteads. This sequence – farmsteads, boothlands and seters – is closely linked to land potential, but of course many boothlands and some seters had the potential to develop into permanent farmsteads. On the old settled areas settlement at first took the form of single farmsteads, but eventually small farmstead groups appeared, hamlets and even villages: there was always a tendency to share limited resources, and eventually a type of field system developed, with farms having intermingled strips, simple versions of what appears in much of Europe as fully developed communal townfields.

In these marginal environments of Sweden we have a glimpse of how settlement may have evolved at an early stage throughout much of western Europe, with simpler hamlets and some sharing of arable lands and meadows preceding the growth of larger nucleations with their more highly organised field systems. There is indeed some evidence from England that the roots of the quiescent systems seen in historic times are to be found in ancient arrangements, often involving a large-scale economic differentiation of space, with arable being concentrated in the regions of better soils, and pastures and woodlands in less favoured environments. For example, twelfth- and thirteenth-century documents from the north of England mention place-names incorporating the term 'shield' or 'shiel', a 'shieling' being an area of summer pasture corresponding to the seters of Sweden. These are usually rather marginal upland settlements. The term 'booth', generally meaning a 'dwelling' or 'house' appears in place-names such as Bothel, Shilbottle and Bolton.

In general, however, the deep roots of English settlement and the presence of very complex mixtures of good, medium and poor land means that the vast majority of settlement names cannot be simply explained in terms of colonisation and changing land-use. Nevertheless, it is possible to give examples of place-names which undoubtedly derive from earlier patterns of economic differentiation. The former presence of economically differentiated settlement is frequently detectable within a region's place-names: thus woodland as indicated by *leah* appearing as *-ley* at the end of hundreds of English settlement names first meant 'glade', then 'clearing' in woodland, while *holt*, a 'wood', *hyrst*, a 'wooded hill', *denn*, a 'woodland pasture for swine' and *wald*, 'forest', all speak of earlier conditions, as do *heath*, implying a 'tract of uncultivated land', *fenn, mersc* and *merise*, terms for varied types of fen and marsh. These words generally have Anglo-Saxon roots and reflect the use by these peoples of lands already much altered by earlier Romano-British and prehistoric peoples (Gelling 1978, 1984). As the juxtaposition of the two diagrams in Figure 6.1, 'Economic differentiation', suggests, it is possible to envisage more complex patterns of regional economic differentiation than those implied by the village–boothland–seter sequence.

Settlement patterns: a summary

The four time–space columns of Figure 6.4 have been considered separately, but in practice if the upper surface represents a block of country perhaps 10 kilometres square, then, in a northern European environment and no doubt more generally in old-settled landscapes, the four columns have to be combined into one to give a glimpse into the complexity of the processes which have generated the pattern which can be observed today (at Tp).

However, generalisations concerning settlement remain tenuous until an attempt is made to apply them in particular contexts. A number of types of settlement pattern are illustrated in Figures 6.5, 6.6 and 6.7, taking specific examples from Europe, Africa and North America. These are all reproduced at the same scale, originally 1:50,000, but here are seen reduced. This demands some concessions concerning what can be shown, while what the source maps actually depict may vary greatly. For example, there is no universal way of showing which isolated buildings are farmsteads and which are not, and while an attempt has been made to create a measure of standardisation, some stylistic variations are inevitable.

Settlement patterns: European cases

Britain: Midland England

Figure 6.5 shows one case from south Warwickshire, in the Midlands of England, where the traditional form of settlement in historic times has been the nucleated village. Large in area, these settlements often show evidence on the ground of both contraction, in the form of earthworks, low banks and ditches where buildings once stood, and comparatively recent expansion in the decades after the Second World War. The former normally results from a reduction in the number of farmsteads, either because of amalgamation or because of an outward movement to new sites, while the latter has been stimulated by the rise of car ownership, which has both helped to retain some elements of the indigenous population and draw in new commuter populations. The place-names – Southam and Napton, Marston and Hardwick – reflect settlement roots in the Anglo-Saxon period – not, it must be stressed, necessarily then taking the form of nucleated villages – while affixes such as Prior's, in this case referring to ownership by the former Coventry Priory, speak of former strong manorial control, a powerful factor in determining the history of each settlement. The second half of the eighteenth century saw dramatic changes in this region's settlements, for following the consolidation and enclosure of the thousands of strips in the communally cultivated open townfields, many farmsteads moved out to new sites at the centres of their newly consolidated ring-fenced holdings, sited where an access road was conveniently placed. This created a pattern of intercalated dispersion to which other isolated dwellings have gradually been added, and may be compared with the changes seen on the Danish island of Samsø between 1819 and 1925 (Figure 6.8). Until the later decades of the eighteenth century these landscapes were dominated by the vast open fields and the term 'champion' (French *champagne*) was applied to them by topographers (Smith, C.T. 1967: 196–211). The end-product of these changes is a landscape of villages with intercalated dispersion.

Bocage landscapes, Flers, Upper Normandy

In sharp contrast (Figure 6.5) the Normandy landscape around Flers is one of essential dispersion, in a region of bocage (Smith, C.T. 1967: 211–18). English terminology would call it a 'woodland' zone, and these are often hilly regions with poorer soils, dominated by hedged enclosures interspersed with surviving woodland tracts. Characteristically they are interfluve regions, whose locations, often remote from main routeways and rather higher terrains, experienced a different sequence of development from regions of better soils in the great river valleys, their terraces and adjacent scarplands. Within woodland regions some of the older settlement cores, often where the churches were located, once possessed limited areas of communal strip cultivation, and the loose clustering of farmsteads present, linked farmstead clusters, as at La Graindorgière (Figure 6.5), which often bear ancient family names, and result from a long process of gradual expansion into reserves of woodland, heathland and rough pasture. The Norman Conquest of England brought William Graindorge to a holding in Yorkshire in 1166, and although we cannot be certain, it is possible that his family did originate in this tiny hamlet bearing his name. The 'nucleations' of this region are of a rather loose texture, congregations of associated farmsteads rather than true villages, originating in the

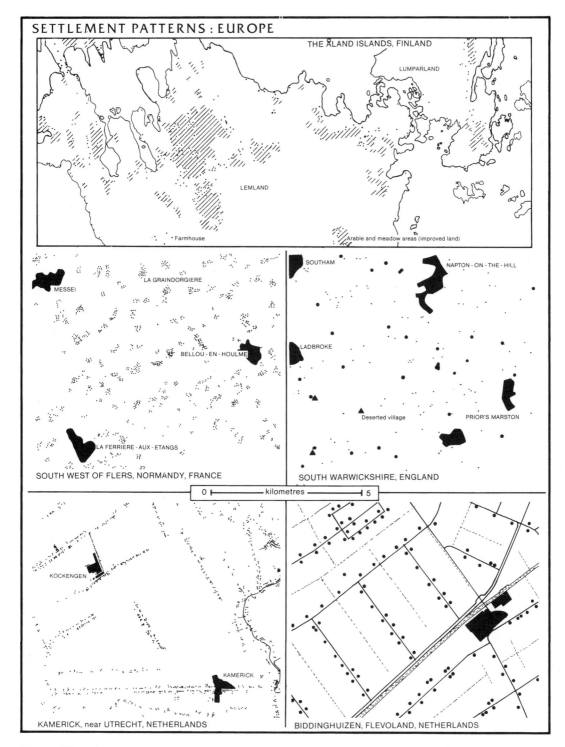

Figure 6.5

division of lands among heirs under the pressures of divided inheritance. As is so often the case, these rural settlements are now losing population, allowing farm consolidation and the use of redundant buildings for holiday accommodation.

To make a more general point, it has to be admitted that the processes which led to emergence of these contrasting landscapes, regions dominated by open communal fields and woodland regions, are not easy to determine. What is increasingly clear is that it is not merely a matter of the former being 'early' settled, and the latter, the woodland zones, being 'late' settled. The contrast, visible in Figure 3.2, was undoubtedly present in England by 1086, for the Domesday Book records large cleared tracts interspersed with more wooded regions, but the ultimate roots of the contrasts may in fact be prehistoric. Land quality, soil type, accessibility to surrounding regions, population levels at critical points in time, type of farming economy, the character of lordship, inheritance practices and the periods of active colonisation have all played a part. Explanation is difficult, but this does not eliminate the uncomfortable fact that such contrasts exist, or that the 'woodland' landscapes of eastern England, with irregular field systems based upon sheep–corn husbandry (Figure 3.8), were the most prosperous and the most populated portions of the realm until the drastic redistributions of people and wealth caused by the Industrial Revolution.

Åland Islands, Sweden/Finland

A fascinating and in many respects 'primitive' landscape is preserved in the Åland Islands of the Baltic (Figure 6.5, upper map). Technically Finnish territory, these islands nevertheless have a substantial Swedish-speaking population and strong bonds with that country, and their physical and cultural characters are highly distinctive. The basal granite of which they are formed, dark red, reddish brown or brown, is still rising after depression by ice, and this is seen along the shorelines where a bright rim of bare rock is visible, in a tideless sea. Intensively scoured by recent glaciation, overlain by fluvio-glacial sands and with small marine clay plains exposed by the rising land – '3 to 5 metres a generation' – this is a fascinating environment (Mead and Jaatinen 1975). In hard winters the islands form part of an ice bridge between Finland and Sweden, but the byword is physical intricacy, for there is a steady gradation from the higher island masses to the *skärgården*, surrounding haloes of rocks exposed or otherwise as the height and state of sea level change in response to pressure and wind. Until the mid-nineteenth century the archipelago was largely self-sufficient, growing small amounts of grain, potatoes and peas upon manured pocket-handkerchief fields, some of which are based upon rather short strips, with ownership intermingled, so that they resemble layouts which may have been the precursors of the great communal strip-field systems of more favoured zones of Europe. These are scattered where arable land can be won and husbanded, and in this context the rigorous control of the difficult physical environment cannot be escaped. The resultant pattern is one of intricate dispersal.

Settlement landscapes: the Netherlands

In the Netherlands the country's historical geography lies close to the surface, for once the firm ground of the glacial deposits which form the east of the country has been left behind one enters a region associated with the deltas of the Rhine and the Maas, the 'low Netherlands', made of marine clays, fluvial sands and peats, a battleground between land and sea, salt and fresh, nature and culture, where centuries of patient reclamation of an ever-changing environment have gradually resulted in a land made secure by technological advance (IDG 1977, 1979). At first clever wooden sluices were used, which let the land waters out at low tide, then windmills were applied to pumping, as well as grinding corn and pounding seeds to extract oils. Finally massive engineering projects linked with sustained water control now use electric and diesel pumps. The Kamerick area reveals patterns of nucleation and dispersion in a region of medieval reclamation. This is curvilinear along the old course of the Rhine (Oude Rijn), a zone of older settlement, and linear where later but still medieval polders – embanked and reclaimed areas – have been established, and where a multiplicity of canals flow into master-drains. Along these embankments are roads, old farmsteads – many almost moated by the vast network of ditches – and more recent dwellings. Near Kamerick itself rather more irregular lines of peat-cutters' cottages can be seen, forming a loose chain to the south of the main east to west dyke axis. The nucleations, now of course representing small service centres with shops, have often arisen where there was sufficient firm ground to establish a church, but historically these are almost incidental to the main business of winning a living from polder and marsh. This is a landscape of infinite subtleties, where the slightest change in level can give advantage or difficulty, and may reflect no more than the depth achieved by medieval peat cuttings. Following an ancient practice along this northern European coast there is a sustained tendency to use any spare materials, from ditches dug, buildings demolished or waste accumulated, to create higher foundations for a new building, creating the *terpen* already discussed (Figure 5.2, pp. 92–3).

The Dutch continue to give their landscapes the feeling of a continually sustained garden, cared for and watched over, thus taming a difficult and often inhospitable environment. Reclamation continues, and Flevoland represents a wholly new polder area, taken from the Ijssel Meer, part of a group reclaimed between 1950 and 1968, with a new town established at Lelystad receiving its first inhabitants in 1984. The area illustrated (Figure 6.5) is generally given over to mixed farming, but also includes a series of rather smaller holdings producing fruit. This contrast is evident in the pattern of farmsteads. The geometric regularity and the absence of ribbon development in this new region is strikingly clear, as is the contrast with the older settled region.

Discussion

A number of general lessons emerge from these European cases. First, the physical nature of the land remains a crucial framework for settlement development, particularly in regions of extremity, i.e. those dominated by poor soils, extreme wetness, cultivable marginality. In regions of intermediate and good soils there are often close responses to the underlying qualities of land and soil, but these are now normally masked by three extremely powerful factors, the cultural alteration of soils, technological advances in all aspects of agriculture and transport, and modern economic pressures. These latter derive both from the economics of agriculture and the demands on land created by other factors which usually emanate from urban populations, such as the need for building land and space for recreation. Second, there is here a paradox: at root, patterns and forms of settlement may owe much to a millennium of work by peasant farmers, and to understand this we must turn to the techniques of archaeology, history and historical geography, but this inherited framework is now, within Europe, being metamorphosed by new pressures, alien to the rural base, whose roots are urban-based: sustained population levels, wealth creation and a cold view of land as an exploitable resource. This is a process which in many ways resembles fossilisation, the replacement of former tissue and bone with something which preserves its outward detail but not its essential qualities. Third, in environments where the contacts between 'traditional' and 'advanced' lifestyles have been rapid rather than gradual, vast tensions are generated. While in Europe these tensions will inevitably emerge in regions until recently dominated and retarded by Communist ideologies, nowhere are they more evident than in the contrast between the developed world and the Third World.

Settlement patterns: African examples

The African cases seen in Figure 6.6 show something of the ways in which indigenous and imposed colonial systems interact. The region around Ungogo, near Kano, Northern Nigeria, is very closely settled, with rural densities of the order of 200–400 per square kilometre. This is made possible by the presence of light, freely draining loams, watered by seasonal rainfall. In the depressions and alongside rivers areas of clay soil are suitable for irrigated agriculture. A measure of historical understanding is also essential. In this zone, round about AD 1000, a series of indigenous Hausa city states appeared, skilled in iron-working and based upon permanent agriculture. Wealth was increased as a result of their location at the southern end of the trans-Saharan caravan routes, which by the fourteenth century introduced the faith of Islam. Northern Nigeria is thus a region where African concepts of land tenure are overlain by Islamic land law. Control by the Emir of Kano seems to have been exercised through local sultanates. By the early nineteenth century these Hausa states were under the control of Fulani rulers, in origin nomadic pastoralists, who consolidated Islamic control in the region. In a context where raiding and warfare were endemic, the result of changing alliances, a series of walled nucleations were created, large villages or agro-towns. Some of these survive, as at Ungogo and Sabon Birni, but others are now almost wholly deserted ruins, integrated within a farmed landscape. Where they survive cultivation is now the most intense nearest to the settlement, sustained by donkey-loads of manure brought from the village, with bush fallow lying further out. Bush and grazing land survive on rocky outcrops and in distinctive strips along village boundaries.

The area illustrated includes four villages or quasi-towns, the scatter of compounds and woodland boundary strips suggesting the mosaic of territories superimposed upon the varied terrain beneath. This is a complex pattern; the compounds are absent where the land is farmed from the nucleations, and in recent decades a tendency towards increased security has encouraged dispersion. Nevertheless, the role of the nucleations as markets and service centres has also increased, focuses for developing transport networks. This creates a curious dichotomy, the nucleations becoming fixed points, with

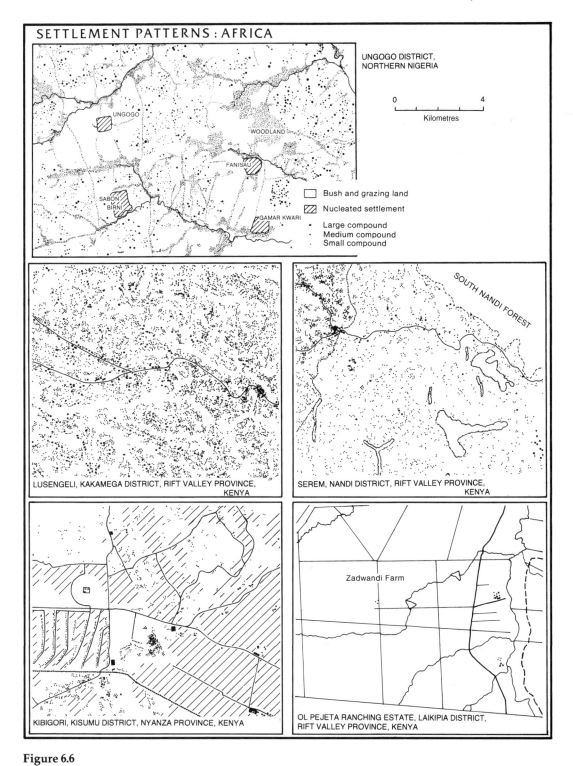

SETTLEMENT PATTERNS : AFRICA

UNGOGO DISTRICT, NORTHERN NIGERIA

0 ___ 4 Kilometres

Bush and grazing land
Nucleated settlement
Large compound
Medium compound
Small compound

LUSENGELI, KAKAMEGA DISTRICT, RIFT VALLEY PROVINCE, KENYA

SEREM, NANDI DISTRICT, RIFT VALLEY PROVINCE, KENYA

SOUTH NANDI FOREST

KIBIGORI, KISUMU DISTRICT, NYANZA PROVINCE, KENYA

OL PEJETA RANCHING ESTATE, LAIKIPIA DISTRICT, RIFT VALLEY PROVINCE, KENYA

Zadwandi Farm

Figure 6.6

mud-brick compounds packed between narrow streets, while, as in much of Africa, purely rural compounds show a tendency to shift as occasion demands. The pattern incorporates stable and less stable settlement elements, and perhaps reflects a stage passed through by European systems before quiescence was achieved (Figure 6.2). This is an African landscape whose known historic antecedents have a peculiar fascination for anyone versed in the historic landscapes of Europe.

The Highlands of Kenya

The remaining four examples of African settlement are taken from Kenya and owe much to discussions with W.T.W. Morgan. In this area, over 5,000 feet above sea level and hence more temperate, between the indigenous tribes there were to be found broad areas of no-man's-land, areas put aside by the British colonial rulers as suitable for white settlers. The railway was crucial to opening up this area and for the export of commercial crops. The scattered dots represent 'houses', compounds in the case of the African settlements, but these contrast markedly with the vast Chemelil Sugar Estate in Kibigori, under European control, and to the west a series of Indian smallholdings also producing some sugar, the cane fields represented by diagonal shading. The larger structures are a 'Chief's camp', 'labour lines' and a village adjacent to a market, although other markets do appear, often by stations, but virtually independent of any sustaining nucleation. In European minds markets and towns are inextricably linked, but in fact there is no reason in rural areas, particularly where the local population is dispersed, why trading cannot take place in an open space. In fact, through time, rulers, administrators and landowners have sought to control and tax trade, and this in part accounts for its link with ancient urban foundations in Europe, but informal and semi-formal trade locations, sometimes with a deeper and more ancient social or religious significance, are a universal feature of settlement patterns. At the present time 'car boot sales' are a distinctive feature of the English Sunday landscape, a return to informal trading in rural locations, free of many urban constraints and restrictions. A more extreme form of colonial settlement is seen in the Ol Pejeta area, on the north side of Mount Kenya, a dry landscape of flat lava flows. This was once part of the territory of the Masai, nomadic cattle herders, who used it as part of their annual cycle of migration in search of grass and water. Water is crucial, in part brought by pipe from Mount Kenya itself, in part provided from streams and boreholes. Overall this is a zone of very low stocking densities and even lower population densities, with isolated ranch farmsteads linked by limited all-weather roads and tracks which are seasonally impassable; grass airstrips are common.

These two settlement patterns contrast with regions settled by African farming communities. The Kaimosi area, Kakamega District, is one of the most densely populated rural settlement zones of East Africa, with between 1,000 and 1,500 persons per square kilometre, with levels often lying closer to the former than the latter. There is a concentration of higher-status rectangular buildings along the roads, and a tendency for services – chief's office, dispensary, markets, churches and schools – to congregate there but the subtle variations within the overall pattern are because settlement avoids the stream valleys. In less densely settled areas these tend to carry forest and are normally very wet in the rainy season, being floored with a deep uniform layer of clay, which cracks in the dry season, tearing roots apart. Add to this a tendency for insect pests to favour such zones, and the reasons why they repel human settlements are clear. In sharp contrast, and no more than 6 miles or so to the south-east the density drops dramatically, but this does not involve different terrains, merely the crossing of a provincial boundary, between Kakamega District, occupied by a tribe of Bantu cultivators – the Kakamega – and a tribe of Nilo-Hamitic cattle herders in Nandi Province – the Nandi. Between these two lie deep cultural barriers, affecting diet, use of land, marriage practices, fertility levels. Thus the warriors of the cattle herders tend to live apart, their sexual experiences being limited to contact with pre-pubescent girls, so that the overall birth rate is lower than that of the agriculturalists. Add to these contrasts the powerful factor of colonial control, discouraging and preventing warfare and territorial expansion by the more populous group, and a stark contrast in settlement emerges. The zone wholly lacking settlement apart from a slight nibbling at the edges is a forest reserve, and the two maps show well the problems of preserving these in the face of such pressures on land.

Discussion

These examples show the role of colonial forces in moulding settlement in Africa. In some areas distinctive new patterns were thus imposed, wholly modifying the direction and character of traditional arrangements; but in others the effects were more subtle. Nevertheless, the example from Nigeria carries the argument further, for it is a powerful reminder that neither Africa nor any other part of the world, other than a very few isolated areas such as pre-contact interior New Guinea, has ever been free of an interaction between endogenous and exogenous forces, whose interaction and balance, coming together, generate new forms and new patterns.

Settlement patterns: the Americas

Eastern Iowa (Figure 6.7) is a rolling till plain dissected by a fine network of dendritic streams flowing a hundred or so feet below the general level of the country. Once a landscape of open windswept grasslands, the area is now fertile and widely cultivated, and superimposed over the relief is a regular grid of roads and boundaries which delimit 6-mile-square townships, each further divided into mile-square sections. This system of land division, having roots as far back as classical antiquity, was applied during the opening of the Middle West to farm settlement. After 1862 each homesteader could claim a quarter section of public domain land (160 acres) by the simple procedure of filing a claim and paying a small fee. After five years of occupation, full legal title was conferred (Conzen 1990). As settlement moved westwards many thousands of acres were so divided. Such organised, rigidly geometric regularity is characteristic of land-taking in colonising contexts. An interesting point about this example is that the irregular terrain has meant that individual farmsteads, while of necessity responding to the basic geometry of the imposed land system and normally located adjacent to roads, have also responded to the underlying terrain, so that the irregularities in the distribution can be related to the choice of particular site characteristics by individuals. Incipient nucleations have appeared along major roads, particularly where an earlier trail cuts across the grid system.

The case of San Julian colony in Bolivian Amazonia represents an attempt to move away from holdings laid out in a 'piano key' framework, which tends to lead to land

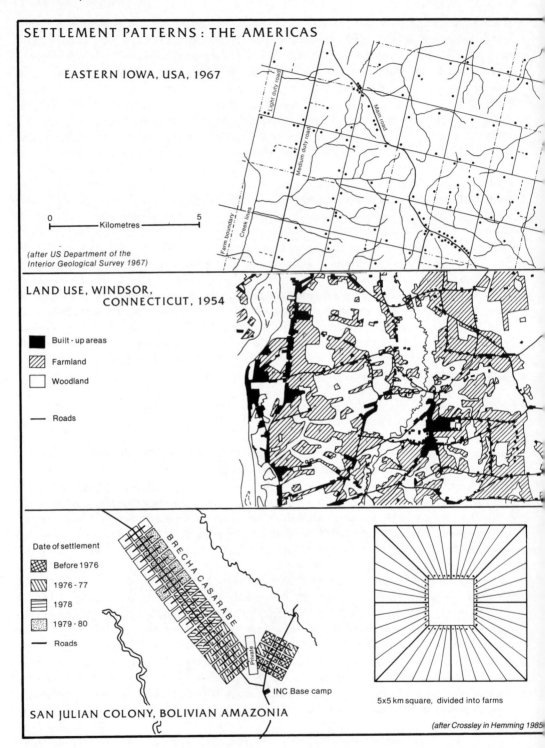

SETTLEMENT PATTERNS : THE AMERICAS

EASTERN IOWA, USA, 1967

light duty road

Medium duty road

Main road

0 — Kilometres — 5

Farm boundary

Creek lines

*(after US Department of the
Interior Geological Survey 1967)*

LAND USE, WINDSOR,
            CONNECTICUT, 1954

■ Built - up areas

▨ Farmland

□ Woodland

— Roads

Date of settlement

▨ Before 1976

▨ 1976 - 77

▤ 1978

▨ 1979 - 80

— Roads

B R E C H A   C A S A R A B E

Private

INC Base camp

SAN JULIAN COLONY, BOLIVIAN AMAZONIA

5x5 km square, divided into farms

(after Crossley in Hemming 1985)

Figure 6.7

abandonment because of the physical and social isolation of the settlers. The National Colonization Institute (INC) began to settle the area in 1968, attempting to incorporate an arrangement of holdings based upon a 'nucleated village'. Each farmstead comprises a radial holding focusing upon a 2-hectare communal plot, the total cluster comprising forty holdings. The construction of roads and wells was seen as essential to the support system, although maintenance of the former proved to be a problem. This layout may be compared with that of Moshav Nahalal seen in Figure 5.2 and with the pattern of eastern Iowa also seen in Figure 6.7. There are in fact powerful links between this Bolivian case and the homesteads of the Middle West of the USA: the North American elements of the United Churches Committee, a body comprising North American and Bolivian members of Methodist, Catholic and Mennonite churches, are themselves the children and grandchildren of small farmers of the Middle West and have a desire, either conscious or subconscious, to replicate in Bolivia a way of life that they found satisfying in their homeland (Crossley in Hemming 1985: 182). There are two problems: first, these Amazonian homesteads are unable to bear the sustained cropping found in the Middle West of the USA and the Bolivian example appears in a context where shifting cultivation is the only proven way of achieving adequate yields. Second, colonisation in the USA was linked to an expanding economy which quickly integrated the new ventures into a broader national and world economy via railway routes and sea links. In Bolivia the proportion of the population still growing their own food is sufficiently great for the ratio of producers to consumers to act as a barrier.

Discussion

Of course, by no means all New World settlement reflects recent colonial origins: the example included in Figure 6.7 from Connecticut, one of the older colonies, reveals a matured and intricate landscape, superficially indistinguishable from long-settled areas of Europe. Nevertheless, in all of these cases, the visible settlement patterns can only be understood in terms of a historical dimension. Between the detail of individual settlements and the generalised pictures provided by maps of patterns there are many possible scales of analysis, and the remaining examples in this chapter are chosen to illustrate both varied scales of analysis.

Settlement patterns: historical dimensions

The island of Samsø, Denmark, was mapped in the early decades of the nineteenth century soon after the scattered strip holdings of the communal townfields had been consolidated (Figure 6.8). At that time, apart from a few large single farmsteads in the hands of landowners, the general mass of the population lived in large nucleated villages, generally rather irregular agglomerations, with occasional regular planned additions giving them a composite structure. Many were arranged around ponds, once important for water supply in this landscape of low rolling morainic hills. All were set back from the coast, perhaps a precaution against piracy. It is probably fair to set alongside this map some work done on the island of Fyn, another member of the Danish archipelago, where in the village of Ronninge and other nucleations small excavations were undertaken to recover pottery samples. These finds could then be dated and mapped, and suggest both

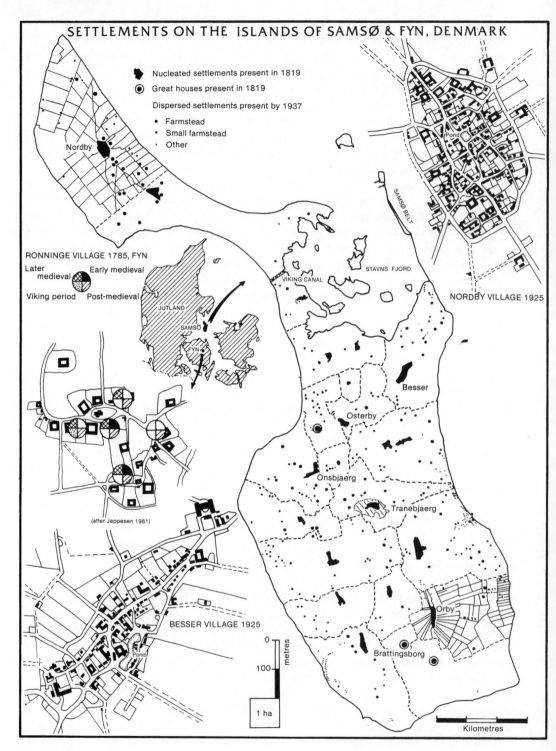

Figure 6.8

a long period of site occupation reaching back to Viking times and also accretive growth from smaller nuclei, for the earliest, Viking, pottery tends to be found concentrated into one part of each village site. The Vikings in fact cut a canal across the narrow beach ridge linking the north of Samsø to the south so that ships could be moved from one side to the other without having to beat round in the face of unfavourable winds. This gave the islanders control of the sea passages to the east and west, and indicates one reason for avoiding coastal locations: not all seafarers were law-abiding or willing to be taxed.

The late eighteenth- and early nineteenth-century process of consolidating the strip fields led to the creation of a radially organised or stellate pattern of farm boundaries, as at Orby, an arrangement that was designed to retain the farmsteads in the village. This did nothing to cure the key problem the farmers had when they possessed scattered strips, namely the distance they had to travel to cultivate the outlying fields. In contrast, at Nordby the consolidation was into blocks, which created an incentive for the farmsteads to move out of the village. This factor, cost–distance, was a prime cause of the outward movement of many farmsteads in the years between 1819 and 1925. The development of such intercalated dispersion after the consolidation and/or enclosure of strip fields is a common feature of the village landscapes of England, indeed of the whole of Europe, but two things must be stressed. First, in England consolidation was invariably associated with a physical process of enclosure, the hedging or walling of the new farms, but in Denmark, the boundaries remain 'open', marked only by a path, a strip of grass, or even merely a change in cropping. Second, it is tempting to call this dispersion 'secondary', and in some senses it is, for it undoubtedly succeeds the nucleations, but in their turn the nucleations may, as the Fyn evidence hints, have followed an earlier phase of dispersion. Terms must be selected so as to be independent of time and place, and in this case 'intercalated dispersion' is both adequate and neutral.

At the scale of the Samsø example it is possible to show every house, but where dispersion is present there are vast problems in examining both nucleation and dispersion over a large area. As Thomas (in Carter 1989: Chapter 8.5) stresses, Welsh medieval society, with a land tenure system which involved division of the land between heirs after the death of a landowner, was associated with an economic system biased towards using the great pastoral reserves of the country. Early dispersion became deeply rooted and Dwyran Esgob (Figure 6.9) in Anglesey reveals both the fragmentation of farms which may go with such a scatter and the association between this fragmentation and the need to obtain a foothold on the best arable soils. Nevertheless, even in this region dominated by dispersion, elemental nucleations have been present for many centuries, and in this case the map (Figure 6.9) involves more than morphology or population size, for the larger symbols, 'villages', denote settlements with a church, public house, post office and school, the 'hamlets' lacking one or more of these. Many of these clusters are of ancient foundation, and reflect factors of political history and of land quality in that they appear in zones of good- or medium-quality soils, much of which are now under grass, but which, under different economic conditions, were suitable for arable. There appears to be a link with the thrust of English control in the twelfth century along the coast of Glamorganshire and into Pembrokeshire, and into the eastern and northern borderlands of the principality. Many of these simple nuclei, and Llannefydd (Figure 6.9) is a good example, are now of enhanced importance because they have gradually accreted more recent service functions to an older core.

For Wales, dispersion has been laboriously mapped by means of the time-consuming

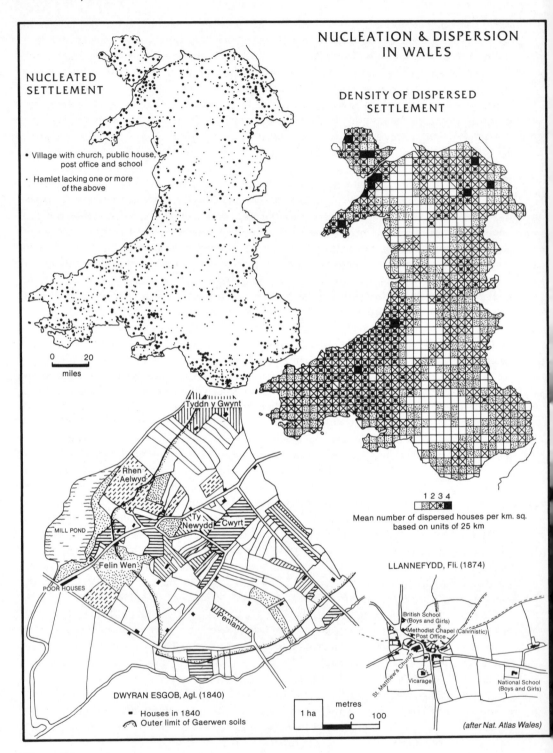

Figure 6.9

technique of counting the number of dispersed houses in blocks 25 by 25 kilometres, and then averaging on a per square kilometre basis (Thomas in Carter 1989: Chapter 8.5). The map of dispersion is wholly inseparable from that of nucleations: they must be read together. During the eighteenth and nineteenth centuries a rising population and increasing markets within the United Kingdom for products such as slate stimulated upland colonisation, with the settlement frontier being pushed higher along the mountain slopes. Holdings that had formerly been only intermittently occupied as adjuncts, dependencies, of lowland farms now became the nuclei of independent units, although their earlier function – as shielings or outlying pastures – is often indicated by the retention of the name *hafod*. The decline of hill farming in the more exposed and remote localities has, in this century, left many isolated steadings derelict, while others have become second homes or retirement cottages, giving (so we are told in the holiday brochures) 'simpler pleasures ... enjoyed beneath the thatch ... where Wesley once preached', with 'sheep as nearest neighbours', and 'a cosy cottage ... sandwiched between the scenery and the beach...'. Elsewhere, of course, a whole rash of scattered houses have been selectively reinforced by the construction of modern residences, often extending older nuclei or infilling once empty coastlines with the ephemeral settlements of caravan and camping sites, also subject to seasonal occupations.

Conclusion

What general lessons emerge from analysis of settlement patterns? First, while in all contexts the qualities of the terrain represent a fundamental framework which must be taken into account, it is only within the more marginal environments, where the constraints upon the character and diversity of land-use are most felt, that physical deterministic explanations become dominant. Second, societies impose upon this base a territoriality which, while at first expressed purely as simple spatial economic variations dependent upon terrain, i.e. using land for which it is best fitted at a simple level of technology, gradually becomes increasingly subtle as land usage, technology and landownership mature. In regions where soil quality is successfully high, and where political conditions are favourable, the fundamental pattern stabilises. Nevertheless, individual elements of the pattern, i.e. the forms, will either stabilise themselves or, more usually, experience expansion or contraction. In this way the character of the overall pattern and the proportions of nucleation and dispersion present gradually adjust to the new circumstances. It follows that to understand and explain structural changes in patterns, awareness of the particular trajectories of individual places is crucial.

CHAPTER 7

Settlement systems: a world view

To provide a composite picture of global rural settlement is no easy matter, yet the detail of earlier discussions undoubtedly requires a broader context, a structure, which synthesises some at least of the picture. To achieve this, the final chapter will take three steps: settlement as the physical reflection of the social organisation of space will be treated historically in a first section; next, world socio-economic zones present by 1500 are assessed, together with the general thrust of subsequent changes; finally a classification of global settlement systems is given. Three final maps show settlement variations at the scale of the sub-continent, i.e. Europe, India and Africa.

The social organisation of space

Figure 7.1 is based upon a study of social and economic change in Europe between prehistory and the industrial era by Dodgshon (1987). It provides a useful starting point for discussion. All advanced animals have an organised relationship with the natural environment; the size and age structure of the group, the size and character of the territory and the food supply it offers are all important. Human beings are unusual because they organise space in conscious, deliberate and planned ways, and this process has, by means of that ultimate magic, the written record, helped human societies to become highly structured or bureaucratised. At root, of course, lies basic experiential space (Figure 7.1, Level I), deeply buried in our psychological make-up, consisting of two elements: enclosed space, with us curled up and protected – a memory we all retain of the safe, warm womb – symbolically expressed by the cave, home and safety, and through-passage space, the space of the natural environment, the treetops of our ultimate ancestry, where movement, although it may eventually be constrained by mountain, desert or sea, is possible in most directions. Through-passage is something to be tested, pushed against and explored.

Beyond these beginnings, culture-space relationships are here categorised by four levels (Figure 7.1, II–V), which leap from the world as perceived by simple egalitarian tribes to the modern scene. Each level, each temporal stage, possessed distinctive

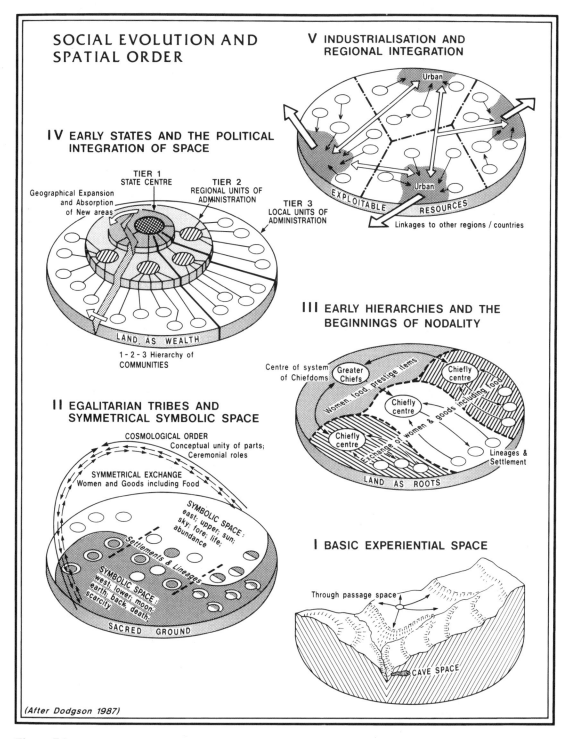

Figure 7.1

settlement characteristics, and as late as the nineteenth century elements survived which were representative of each level. We will take each in turn.

Egalitarian tribes

The earliest European societies consisted of small groups supporting themselves by hunting and gathering: once farming was introduced the communities which developed utilised a wide range of natural resources and sustained local links between groups by means of formal exchanges of food and that most precious of precious commodities in a primitive society, women. In effect this was a grant to another group of a capacity to bear their children. Such small groups, clans, appear to have imagined space as divided into two major parts (Figure 7.1, Level II), one symbolised by words such as east, sun, upper and sky, the other by words such as west, moon, lower and earth. These two dimensions of a divided and chaotic cosmos were brought together within myths, stories expressing inner truths, and this togetherness was ensured, enacted and cemented by means of group ceremonials.

Settlements were part of this perception, with individual entities thought of as being associated with a particular cosmological dimension. Such ideas have been documented by anthropologists working among near contemporary traditional societies. Indeed, they may underlie many aspects of settlement we now see as functional and prosaic. To reiterate one example given earlier, the many northern English and Scandinavian village plans which have a strong east-to-west or north-to-south orientation derive this not from 'natural' site factors but from ancient, deep-seated pre-Christian folk beliefs. What people *believed* was as much part of the societies and settlements they created as were the tool-kits they had at their disposal and the type of economy they evolved. The crucial step of plant and animal domestication is as likely to derive from ritual practices as from purely 'economic' needs.

Tribal chiefdoms and early hierarchical societies

The eventual concentration of power and prestige, both secular and religious, in the hands of individuals and families drew the individual settlements of clan lineages into groupings (Figure 7.1, Level III); the chief received from the places under his control payments in the form of food, women and other goods, and the same things, plus other prestige items, were exchanged between chiefs of adjacent groups. This concentration of power together with the increasing militarisation of society led to the appearance of great paramount chiefs and eventually kings. At this level prestige trade goods become more important, luxuries such as high-quality metalwork, fine textiles, rare skins and hunting dogs, exotic items imported from other regions and as gifts from other rulers. Space was organised into the territories of a series of tribal groups, some of which recognised a paramount chief or king. The relationships between these could vary from wholly cordial to wholly hostile. In Europe this system was in part swept away and in part formalised by the military and economic colonisation of an advanced Mediterranean culture, Rome, expanding to obtain supplies of slave labour by way of conquest and raw materials by way of tribute and trade (Figure 3.6).

Feudalism and early states: traditional societies

Feudalism was a system through which the energies and resources of the land were harnessed to support a military and ecclesiastical elite in the centuries following the collapse of the Roman order in western Europe. Similar autocratically organised societies are known worldwide, for instance in Japan. 'Those who fight, those who pray, and those who work to keep the rest alive' was a cynical medieval appraisal. In Europe control was effectively sustained through the two interlocking frameworks of state and church (Figure 7.1, Level IV). The state was organised hierarchically, with the power of the king and the noble class linked to important regions, core areas of what were eventually to become the national states of Europe, the south-east of England, the Plain of Uppsala in Sweden, the Paris Basin and the middle and lower Danube Valley (Figure 3.5). Regional units of administration focused upon other royal or baronial centres, in both core and peripheral zones, while at the base lay the settlements of the local communities whose rents, renders and labour supported the feudal superstructure.

A bonding mosaic, tied directly to the land itself, comprised estates in land, held by the barons, both laymen and churchmen, on which the peasantry toiled. The resulting pattern is so intricate, so complex, that when we find it documented at the high point of the medieval centuries, say between 1100 and 1300, it is abundantly clear that it had ancient, diverse and fragmented roots. There was indeed much internal diversity. The great fertile river plains of Europe tended to be occupied by a peasantry upon whom the demands of the lords rested heavily, but in the forests, marshes and along state frontiers greater personal freedom was to be found, a simple way of attracting and holding colonists, of both lordly and peasant status.

Into this society were intruded other social groups, precursors of a later system, comprising town-dwellers and traders, merchants and craftsmen, city dwellers and seamen, who established their own networks within the patterns of feudalism, trade routes, points of exchange and defended emporia – trading centres (Hodges 1982: 45–65) – all nominally regulated and controlled by the crown or the landowning nobility, but ultimately proving more enduring than feudalism. Through these networks industrial capitalism eventually grew from the dissolution of this feudal ordering of space. The economic historian Rostow usefully identified several stages in the development of national economies (Rostow 1960), but it is worth quoting here his proviso: 'I cannot emphasise too strongly at the outset, that the stages-of-growth are an arbitrary and limited way of looking at the sequence of modern history' (Rostow 1960: 1). Essentially, Rostow's stages begin with what he terms 'the traditional society', all types of social order present *before* the beginnings of industrialisation. Creating cross-linkages between differing systems of stage classification is always difficult, but in Europe Rostow's first stage, 'The Traditional Society', broadly corresponds to Level IV of Figure 7.1. Of course, these feudal/traditional societies were never static, but their central character lay in the fact that there were severe technological and conceptual limitations on the levels of output attainable per head of population. The output of farming was limited by prevailing husbandry practices, while the output of craft workers was limited by the productivity of their labour and the available machinery. The rational patterns of improvement deriving from 'scientific and modern ways of thinking' were not generally present. The sphere of the mind was not necessarily smaller, for many great churchmen and nobles and their administrative staff were widely travelled and highly literate men; but it was different.

Industrialisation and regional integration

In fact Level V of Figure 7.1 reveals the temporal distortion present by compressing into one plate no less than four of the phases of the development of industrial societies recognised by Rostow, i.e. the preconditions for take-off, take-off to self-sustaining growth, the drive to maturity, and high mass consumption (Rostow 1960). It is evident that each of these levels is associated with distinctive settlement characteristics, reflecting the transition from ephemeral to permanent settlement discussed in Chapter 2, from subsistence economies with limited trade to the integrated world economic system achieved by Level V. These differing views of space, from the sacred grounds of Level II, via land as roots and land as wealth, to land as an exploitable resource of Level V, encapsulate the changing socio-economic dimensions within which the evolution of the present settlement systems must be set. We are, however, left with a further question: can the diverse socio-economic systems modelled as a through-time sequence in Figure 7.1 be translated into spatially differentiated zones at a global level?

Early world economic systems

The upper portion of Figure 7.2 shows a world pattern of dominant economies in about 1500 and superimposes upon this a generalised distribution of major hearths of plant and animal cultivation. It thus creates a broad picture of the types of rural system which lie at the roots of contemporary landscapes. Five categories are identified, with agriculture divided into two, that based upon the plough, using the energies of draught animals to till the soil, and that based upon hand cultivation, using the hoe, the digging stick or the foot-plough – a variety of spade. This division is also closely linked with a great divide in types of crop, the plough being associated with the cultivation of hard grains, wheat, barley and rice, the other implements with crops which reproduce vegetatively, i.e. from a severed portion, such as the yam, cassava and the potato. There is a deep cultural bias in Europeans towards seeing plough agriculture as more 'sophisticated', but in fact the intercropping and multiple techniques of the tropics and sub-tropics, often linked with terracing and careful manuring, allow highly productive subsistence agriculture. A well-run garden has different qualities to an extensive monoculture grain field; one is not 'better' than the other. Ploughs, however, do tend to demand larger and more regular fields than hoe cultivation and operate with systems which use a shorter fallow (Ruthenberg 1980: 70–1, 107).

By AD 1500 a zone of plough agriculture embraced the more habitable portions of the continents of Europe and Asia and extended into northern Africa, with the Sahara and Ethiopian uplands clearly forming cultural barriers, although the former was eventually crossed by a network of trade routes made possible by the domestication of that 'ship of the desert', the dromedary or camel. The plough is associated with the early hard grain production and the domestication of sheep and cattle which developed in the Fertile Crescent of Mesopotamia 8,000–9,000 years ago. Even earlier developments in planting agriculture and the domestication of animals, the pig and the chicken, were taking place in South East Asia. There is much debate concerning the processes by which these domesticates were diffused and welded together in the varied mixed farming systems found throughout the Old World. That the idea of domestication diffused and that some

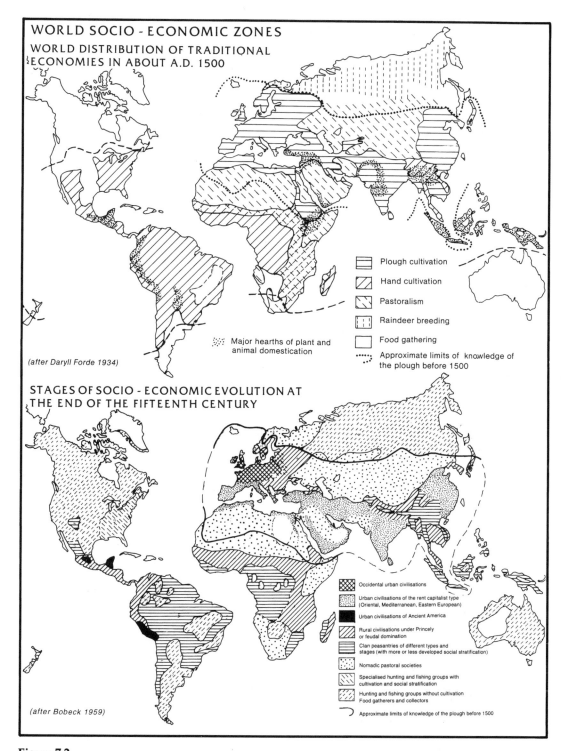

WORLD SOCIO - ECONOMIC ZONES

WORLD DISTRIBUTION OF TRADITIONAL ECONOMIES IN ABOUT A.D. 1500

Plough cultivation

Hand cultivation

Pastoralism

Raindeer breeding

Food gathering

Major hearths of plant and animal domestication

Approximate limits of knowledge of the plough before 1500

(after Daryll Forde 1934)

STAGES OF SOCIO - ECONOMIC EVOLUTION AT THE END OF THE FIFTEENTH CENTURY

Occidental urban civilisations

Urban civilisations of the rent capitalist type (Oriental, Mediterranean, Eastern European)

Urban civilisations of Ancient America

Rural civilisations under Princely or feudal domination

Clan peasantries of different types and stages (with more or less developed social stratification)

Nomadic pastoral societies

Specialised hunting and fishing groups with cultivation and social stratification

Hunting and fishing groups without cultivation Food gatherers and collectors

Approximate limits of knowledge of the plough before 1500

(after Bobeck 1959)

Figure 7.2

of the domesticates, both plants and animals, were spread outwards from hearths of origin is hardly to be doubted, but there were also centres of subsidiary domestication (Sauer 1952; Steensberg 1986; Isaac 1970; Ucko and Dimbleby 1969).

Renfrew has suggested that in Europe the diffusion movements may have comprised an advance of the agricultural frontier into hunter–gatherer territory at a rate of no more than 1 kilometre per year (Renfrew 1987: 124–31, 153–9). This process involved, among other things, the adoption of domesticates by intensive food-gatherer communities already at the threshold of domestication, and the steady and cumulative intaking of new land to replace fields exhausted by cultivation in a context where intensive manuring of more than limited areas was difficult. He postulates that the Indo-European language, the ultimate ancestor of all European tongues (and a proportion of those of India and Asia as well), was an integral element of this diffusion. These movements undoubtedly demanded some sea journeys, not least that across the Aegean (for the Bosporus route was not followed) and the English Channel. Reaching Crete and Greece before about 6500 BC agriculture had spread to the rest of Europe and even the Orkney Islands by about 3500 BC.

Some centuries before the birth of Christ cultivation and cattle-rearing had already diffused throughout the whole of the zone dominated by the plough, and the region's rich forest soils, allowing permanent fields to be sustained by manuring, supported a basic stock of clan peasants finding social focuses in nucleated hamlets and villages (Champion *et al.* 1984: 113–51). Links between the important Middle Eastern hearth of domestication and others in South East and East Asia are suspected and postulated, but in general the story in the tropics is less known, while parts of the New World, particularly in Central and Andean America, were focuses of other hearths of domestication of plants such as maize and potatoes, and of turkeys, guinea pigs and the llama. By about AD 1500 versions of this lifestyle based upon simple cultivation without the plough were still widespread, but were more characteristic of tropical and sub-tropical regions where examples survive there to this day. Thus remote New Guinea peoples still cultivate gardens with double paddle spades (Steensberg 1986: 88–109), a type of sophisticated digging stick identical to that used in Denmark in the later prehistoric period.

The first farmers are generally seen as rather egalitarian tribes (Figure 7.1, Level II), but as population and farming expanded and extended social differentiation and the rise of new technologies can be linked to the rise of a more hierarchically organised social order. In the then fertile heartlands of the Middle East and Mediterranean trade, metallurgy and agricultural production sustained by hydraulic techniques were sufficient to allow the rise of city states and eventually militaristic empires (Collins 1975; Ucko *et al.* 1972). Steensberg argues that it was the practical capacity to grow, accumulate and store (for as long as sixty years) the hard grains which was the key stimulus to the development of weights and measures, land surveying, numeracy and literacy, which lies at the root of the rise of all urbanism and civilisation. Hard grain was a form of capital accumulation whose importance is attested by the biblical story of Joseph and the seven years of plenty and the seven years of famine. Less altruistically, these grains could sustain armies and build tribute-paying empires around the earliest city states of Mesopotamia, the Nile and Indus valleys and the Mediterranean basin (Steensberg 1986).

Purely pastoral economies represent a secondary development, a specialist adaptation to the more limited environments of the continental interior plateaux and desert fringes, where indeed they still survive. The most convincing argument for the secondary nature of pastoral nomadism is the fact that not only does their range overlap that of plough agriculturalists, they share the same beasts. Several types can be identified: true nomadism occurs when a population, albeit within a defined territory, moves as a group, with settlements being wholly ephemeral and seasonal. In some contexts this stock movement may take place for only half the year and involve only a portion of the population. True transhumance is a third category, comprising those groups where movement involves only a small number of persons who take the stock to traditional grazing areas for part of the year. Pastoralism is an Old World phenomenon, and has penetrated Africa, particularly along the southern fringes of the Saharan belt and through the highlands of the east into the south-east of the continent, where it has pushed back or in part overlain hoe cultivation (Davies 1973: 26–7). Often there are symbiotic links between the pastoralists and adjacent agricultural communities, for diets of meat, blood, milk and milk products need to be supplemented by traded grain and grain products. In earlier centuries, the mobility of pastoral groups, particularly those of the grazing lands of central Asia, their capacity to move limited worldly goods as packs on animals or in carts, and above all their capacity to use the horse and short bow in military operations, made them much feared by the surrounding settled communities as they were capable of wide conquest and empire building (Piggott 1983).

The social effects of these developments in production can be summarised as follows: in about 10,000 BC, soon after the great ice sheets retreated, 100 per cent of the world population were hunters and food-gatherers who successfully colonised most environments for they were able to draw upon solar energy in the form of plants and animals and had control of stored solar energy in the form of fire. By 4000 BC food production had extended from core hearths between the Balkans and the Irrawaddy and the coastal margins of mainland and peninsular east and South East Asia into peninsular Europe, savanna Africa, and most of east and South East Asia, with apparently independent nascent agriculture along the Pacific side of the Americas from Mexico to Peru (Simmons in Wolman and Fournier 1987: 45–77). By AD 1000 agriculture had spread within South and North America, so that the continent's grasslands, together with Australia, held most of the remaining hunter–gatherer populations, though there were significant numbers of coastal hunters in the Arctic regions. By AD 1500 the world population numbered approximately 350 million of whom only 1 per cent were then hunter–gatherers; only Australia remained untouched by agriculture until the coming of European settlers.

Stages of socio-economic evolution by 1500

Figure 7.2 (lower) is derived from a map compiled in 1959 by the German scholar Hans Bobeck (in Wagner and Mikesell 1962: 218–47). It shows the world distribution of levels of socio-economic development at the end of the fifteenth century, just at that crucial point when Columbus, in 1492, made contact with the tropical portions of the New World. The Vikings had made even earlier contact: their deep-sea trading vessels, capable of moving people, goods and stock over immense distances, had carried colonists to Iceland, southern Greenland and even the coasts of Labrador and Newfoundland, but these activities little affected the mainstreams of European development (Jones, G. 1968: 289–311) and were mere ripples in comparison with the great waves of European world exploration which came in the next century, explorations which nudged England's economy to the preconditions for take-off. This complex distribution can only be understood in terms of developments through time and here only one aspect will be selected for further comment.

Old world agricultural societies

Within the old world areas dominated by plough agriculture Bobeck identified four types of socio-economic system: clan peasantries with more or less developed social stratifications; rural civilisations organised under feudal or autocratic lines; rent capitalist urban areas; and occidental urban civilisations, approaching productive capitalism, industrial society and modern urbanism (Bobeck in Wagner and Mikesell 1962). It will be noted that by 1500 the clan peasantries, survivals of a basic, simple mixed farming economy, with a rather egalitarian social structure based upon family links, have been marginalised within the zone of plough agriculture; they lie on the peripheries, often in difficult, less rewarding environments. In contrast, the feudally or autocratically organised rural societies (Figure 7.1, Level IV) were still to be found in eastern Europe and Russia (where the serfs – unfree cultivators, bound to the land and part and parcel of an estate – were not emancipated until 1861), also in the Indian sub-continent, South East Asia and Japan. These feudal survivals were predominantly rural in character, with groups of peasants remaining under the control of aristocratic families owing, more or less, allegiance to a prince or king. Bobeck's 'rent capitalism' is a broad category, and describes an economic system which clearly derived from what has earlier been termed 'tribal chiefdoms and early hierarchical societies'. It involves a process by which the surplus of the peasant farmers was constantly skimmed off as work or renders in kind or cash or rents (as in feudalism), but it also included the regular advancing of loans and creation of indebtedness. Already present in Mesopotamia as early as about 1700 BC, this system built the pyramids and sustained the military imperialism of Babylon, Assyria, Egypt, Greece and Rome. It created the cities and tombs of China and India, achieving vast cultural flowering on the shoulders of cultivators and craftsmen. All of these societies were urbanised, technologically skilled and administratively advanced, indeed all were literate – the surviving records, buildings, jewellery and bric-à-brac of everyday life attest this clearly. Nevertheless, these societies rarely applied inanimate power to the production of goods, an exception being the Roman use of water power in mills for flour and oil production. This, when linked to aristocratic attitudes which saw production purely in terms of rent returns convertible into military use or luxurious living, rarely in terms of reinvestment for production and sale, meant that take-off to mass consumption was never achieved. Bobeck qualified the term 'rent capitalism' by indicating that he was mapping, for 1500, the urban civilisations of the Orient, the Mediterranean and parts of eastern Europe.

By 1500, the threshold of the 'great age of discovery', northern and western Europe consisted of a series of states, emerging from feudalism and developing vigorous trading links which supported towns and cities ranging from purely local market centres to international emporia, with contacts which extended throughout the limits of the known world. The facilitating break-through was the application of capital, now accumulated in the form of bullion, coin, but also present in less tangible forms, promissory notes and bills of exchange, mere paper, but supported by custom and eventually backed by the authority of state treasuries. Increasing technological competence, linked with an attitude of mind that saw willed change as desirable, gave these European peninsular civilisations an outward impetus. Saying this is to recognise that while the Chinese possessed technological skills that equalled, indeed often exceeded, those of the west, in the centuries after 1500 they looked inwards rather than outwards, possibly because of the large size of the country with its vast land frontiers, its physical, economic and cultural diversity and the presence of remarkably centralised control from an early stage. China also lacked Judaeo-Christian belief patterns, and two attitudes in particular have encouraged European expansion. The first is expressed in the statement that man should 'have dominion over the fish of the sea, and over the fowl of the air, and over every living thing that moveth upon the face of the earth' (Genesis 1: 28), and the second in the exhortations that 'repentance and remission of sins should be preached in his name amongst all nations' (Luke 24: 47) and 'Go ye therefore, and teach all nations, baptizing them in the name of the Father, and of the Son, and of the Holy Ghost' (Matthew 28: 19). Where explorer and trader went, the sword, the cross, and, it must be said, the stake and intolerance of other ideas, soon followed (Las Casas in Carey 1987: 82–4).

Movements, migrations and empires

The latter remarks form an essential preliminary to what was to follow: the upper portion of Figure 7.3a is a reminder of the great changes wrought since 1492, particularly those occasioned by European seaborne explorations of the world, their discovery of 'new' lands, the establishment of trading links and eventually the colonisation of new territories by administrators and settlers. This process was undertaken with the aim of integrating the new territories into the developing world of industrial capitalism and exploiting them for the benefit of Europe. Only the most substantial movements are shown, beginning with that of the Portuguese and Spanish in relatively small numbers to Latin and Central America, followed by a prolonged and accelerating movement of Europeans seeking to take advantage of the lightly populated spaces of the North American continent. Both migrations had disastrous effects upon indigenous peoples, and generated another great movement, the transport westwards of vast numbers of enslaved Africans, men, women and children. The European exploitation of Australia, at first a dumping ground for convicts, was eventually followed by sustained colonisation, while trade links and imperialism sub-divided a weakened Africa and led to European expansion, absorbing India and portions of South East Asia, and only halting finally on the margins of China. Nevertheless, thousands of Indians and Chinese were moved to other regions as 'indentured labour'.

Greed, brutal compulsion and stark need were the driving forces behind these migrations, in the best situations coupled with an optimistic view of the freedoms and opportunities for betterment the new lands might offer. Nevertheless, in many cases the severance from home roots was bitter and savage. There were, of course, many other movements, some over short distances while others were transcontinental, for example Irish and Scots into England, Russians to Siberia and Central Asia, and the movement of millions of displaced persons within Europe following the end of the Second World War. The collapse of the Communist empire in eastern Europe in the late 1980s and the wealth

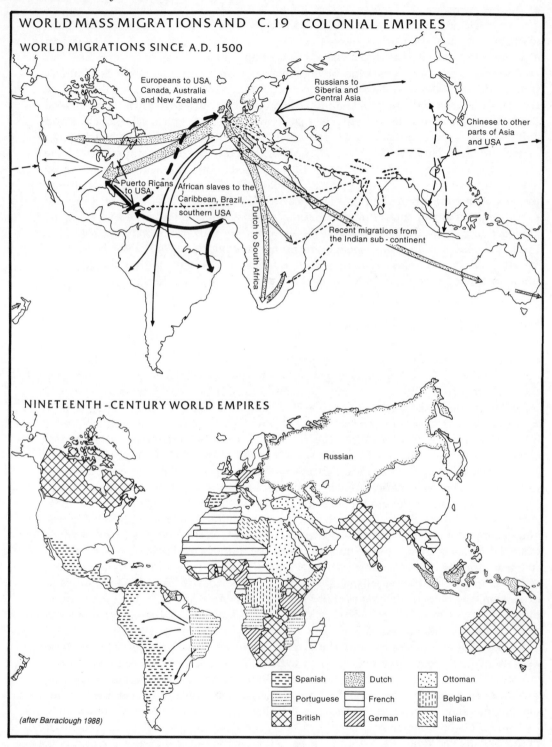

Figure 7.3

discrepancies between Europe and the African countries on the south side of the Mediterranean Sea are currently generating pressures for migration westwards and northwards.

It is in these latter examples that we glimpse something of the driving forces which helped the collapse of the Roman Empire during the fifth century – an earlier 'great age of migrations' – when the 'barbarian' peoples external to the empire were attracted to its visible wealth, seen in buildings, works of art, institutions, lifestyles and patterns of consumption (Cunliffe 1988). They learned about it from visits of envoys, traders, the movements of time-expired soldiers from the legions, seamen and imperial policies of 'buying off' peripheral kings and chiefs. Rather less efficient than television, these channels nevertheless spread information. Thus, and using the words of Taliesin sung between 570 and 580 (and written down much later), the Celtic rulers of northern Britain as much as the wealthy owners of great Roman estates sought

> riches a-plenty
> and gold, gold;
> gold and gift,
> esteem, –
> and estimation ...

(Pennar 1988)

Folk movements have had two marked effects; they ensure an intermixing of races. 'Racial purity' has long been a destructive myth, driven by the fact that human beings seem to need easily definable 'clubs' to which to belong and with whose goals they can identify. While at a world scale it is possible to recognise a broad contrast between the 'black' peoples of Africa south of the Sahara, 'white' peoples of Europe and western Asia, 'yellow' peoples of central and eastern Asia and 'brown' peoples of southern Asia, the sustained intermixing of these groups has been a physical and social fact through all time, creating the present diversity. The numerically large and long-distance movements of the last four centuries have generated a vast potential for further blendings. On the other hand, cultural habits, for instance the practice of marrying within a limited group, have, in many contexts, been a wholly effective barrier to free intermixture, as the hero and heroine of *Romeo and Juliet* always discover. Nevertheless, in memorable words, Brazil remains a country where 'a blond negro can be seen talking Spanish to a red-headed Chinaman' and many parts of the world now show such rich diversity.

All migrations also involve cultural intermixture. Of course, this intermixture may be discouraged, and it is probable that the caste system of India has at root a successive overlayering of migrant peoples, although there are also possible economic causes as well. Culture involves those social dimensions already touched upon above, but also includes the practical aspects of the lifestyle or the *genre de vie*. Between 1492 and the end of the nineteenth century the European nation states established colonial empires (Figure 7.3, lower), and in so doing for ever altered the development trajectory of those regions where they seized or negotiated control, altering societies, economies and, not least, some settlement characteristics.

Settlement systems: a world view

It is in the context of these global developments that world settlement systems have developed and evolved, and it is postulated that five fundamental types can at present be seen (following Thorpe in Wright and Stewart 1972: 106–18):

1   traditional or archaic systems
2   colonial and ideological systems
3   urban systems
4   urbanised region systems
5   ephemeral or encampment systems.

Each will be considered in turn.

Traditional or archaic systems

In the modern world these terms cover everything from the most primitive survivals of hunter–gatherer–farmer communities, to purely pastoral communities and to the farmers of many rural regions. Essentially these are inherited settlement systems, brought to the present after long centuries of sustained development. These systems were in general successful, depending upon mixed farming traditions to sustain permanent settlements, either dispersed single farmsteads or farmsteads concentrated in nucleations. In all their rich diversity they form a substratum of settlement throughout the agricultural landscapes of the world. Many of their characteristic features were already present by 1500, but they have subsequently been subjected to many and varied pressures for change. In all these systems territories were organised hierarchically, with larger, more complex settlements, either rural or urban, being the focuses of dependencies lower down the rank. Towns were brought into existence to serve rural areas, acting as administrative focuses, as exchange points within the system, and as focuses for external trade with other regions. Throughout the developed regions of north-western Europe, technological changes and the dramatic decline in the farm labour force since 1500 has left the physical elements of this system as cultural fossils, functionally severed from their origins. Nevertheless, the inherited frames of patterns and forms are adapted to varied new functions, and the original agricultural core may have craft elements, industrial elements or, eventually, service and commuter/dormitory elements grafted on in varied combinations. This generalisation is easier to grasp when applied to the focal points of hamlets and villages, but is equally applicable to areas of dispersed settlement, where for instance the intrusion of a craft industry could intensify the scatter of dwellings. Defoe, writing in the early eighteenth century, described his approach to Halifax, Yorkshire, as follows:

> But now I must observe to you, that after having passed the second hill, and come down into the valley again, and so still the nearer we came to Halifax, we found the houses thicker, and the villages greater in every bottom; and not only so, but the sides of the hills, which were very steep every way, were spread with houses, and that very thick; for the land being divided into small enclosures, that is to say, from two acres to six or seven acres each, seldom more; every three or four pieces of land had a house belonging to it.

Then I began to perceive the reason and nature of the thing, and found that this division of the land into small pieces, and the scattering of dwellings, was occasioned by, and done for the convenience of the business which the people were generally employed in … this business is the clothing trade.

(Rogers 1971: 491–3)

To a basic landscape of scattered hill farms, the cloth industry, in the late seventeenth century still a domestic craft, had brought the possibility of settlement intensification.

As has been emphasised, traditional systems are, through time, subjected to both internal and external pressures for change, and this leads to either degenerate or regenerate types. In the former, pressures such as overpopulation, excessive sub-division of holdings, outward migration of young people, desertification, the abandonment of traditional values, sustained guerrilla activity, or excessive emphasis upon cash crops – to name only some factors – destroy the centuries-old balances which were the hallmark of successful traditional arrangements. In the latter, traditional arrangements absorb elements of change, such as new farming techniques and crops, new technology and different social values, and achieve a successful blending of old and new elements. In both cases elements of archaic settlement forms and patterns may be sustained, but in the former neglect and decay become dominant features, while in the latter rebuilding and adaptations to the new order are characteristic, often linked to land reform, involving the restructuring or redistribution of holdings.

Colonial and ideological systems

The imposition by an external power of forms and patterns derived from other cultures creates very diverse settlement characteristics; thus, exogenetic forces, for example colonial factors such as the introduction of new settlers derived from European homelands, carried English village plans to New England, German village plans to Australia, and, in general, idealised, often grid-based rural layouts, both nucleated and dispersed, to the Americas, Africa and many parts of Asia. These imposed developments often contrast sharply with the traditional forms and patterns of indigenous peoples. Plantations and large farm landscapes, resulting in relatively regular settlement forms and patterns and in geometric field and farm shapes, reflect the new landholding arrangements of colonial powers. Nowhere is this more apparent than in the Middle West of America, but other variants are at present emerging in regions such as the Amazon basin (Figure 6.7). Regenerate transitional systems, derived from an existing traditional base, may of course owe much to a colonial presence or colonial influence. In the former USSR enforced collectivisation of the peasantry, linked with the devastations created by ideological inflexibility and those wrought by two world wars, led to the imposition of ideologically conceived new frameworks upon older arrangements without the buffer of any transitional phase. In the short term the results were often catastrophic.

Urban systems

Two scales of urban system can be identified: first, simple urban systems exist where a single town possesses an industrial or service base which serves more than the surrounding rural region, for example settlements where an ancient fair drew merchants from a wider area, or a great imperial entrepôt such as Singapore and Hong Kong. Of course, these places also possessed purely local functions, but such towns are distinctive because they have important characteristics which are relatively independent of their immediate surroundings. City regional systems evolve in contexts where a single urban nucleus develops to such a degree as to affect large areas of the surrounding countryside, far beyond the limits of the continuously built-up area. Between the sixteenth and the eighteenth centuries larger Dutch towns developed rural urban fringes characterised by intensive horticulture, small country houses and summerhouses (Haarten in Nitz 1987: 317–28). By the end of the seventeenth century London had grown to the extent of accommodating 5 per cent of England's population, rising to 12 per cent by 1700 (Dodgshon and Butlin 1990: 162), and dominated the regional economies of much of the south-east; for example, its economic influence affected the cattle trade of Wales. Of course, its social influence as the focus of court life and political power had already been present for many centuries. In both of these cases the developing internal structures can be described and explained using the classic models of Burgess, Hoyt, and Harris and Ullman (OU 1970: 349). However, it is important to recognise the primacy of city-centred political and economic control within the modern world, operating at both national and international scales, with a growing complexity of links between the urban congregations, which may often be conurbations, involving several historically identifiable urban nuclei as well as the surrounding rural areas. Most great cities now draw finance, sustenance, raw materials and entertainment from the world. The city is no longer a single, simple entity, if indeed it ever was.

Urbanised region systems

In these, the most complex settlement systems which have yet arisen, all elements of settlement, rural, industrial and urban, are drawn together into a multi-nodal arrangement, classically seen in conurbations and urbanised regions such as the Randstadt, the Ruhr and the eastern seaboard of the USA, but also in vast aggregations such as Mexico City, the antithesis of what urban life should be.

Ephemeral or encampment systems

Finally, it is necessary to identify a special group of short-lived settlements – mining communities, camps of navvies or other workers, military encampments, illegal concentrations of squatters or political refugees – which are imposed on rather than integrated within other adjacent settlement systems. In this category may also be included modern recreational settlements, including some of the 'holiday villa' complexes of the Mediterranean and the 'holiday villages' of the temperate zone, for instance the summerhouses of Scandinavia and even the caravan parks and fields which fill with tents during the brief months of summer. These latter may be transitory in character, but through these categories is a transition, from wholly ephemeral encampments to settlements which form the basis for permanent occupation. All of these types are intruded into countrysides, although few have other than parasitic links to the surrounding rural area, exploiting its qualities in some way, either for the basic needs of shelter, fuel and food, or for recreational purposes. Ephemeral systems have long been an integral part of the wholly nomadic way of life, but modern encampment systems are often a symptom of instability – here one might cite the example of refugee camps throughout the world.

In theory, these varied systems offer the potential of a world map, but in practice time, varied evidence and scale difficulties make this difficult to achieve. However, retaining focus on the specific theme of rural settlements, this chapter concludes with examples of settlement systems mapped at a continental or sub-continental scale.

Continental and sub-continental scales of analysis

The experience of Europe, rich in landscape survivals, documentary sources and work by archaeologists and historians, reveals that the qualities of the forms and patterns when mapped and viewed at this scale do indeed reflect 1,000 and more years of sustained development, destruction and adaptation. While North America may in this context be excluded, perhaps with tongue in cheek, as a mere 'developing country', i.e. one with a recently imposed colonial settlement system, there seem to be no good reasons not to assume that the generalised maps of African and Indian rural settlement types reflect similar roots. These two examples were selected on the purely pragmatic grounds that maps of rural settlement at the sub-continental scale are available, but they represent interesting counterpoints.

Rural settlement in northern Europe

Figure 7.4 is based upon work by many scholars (Schröder and Schwarz 1978: inset map; Roberts 1990: 51–72). This distribution undoubtedly reflects several thousand years of sustained development and change since the arrival of the first farming communities, and includes areas that were brought firmly within the Roman Empire (France, the Low Countries, and England and Wales) and those outside (Scotland and Ireland, Scandinavia, eastern Europe), with Germany forming a complex border zone, divided by a frontier, the *limes*, in effect what in later English history would be termed a 'march' zone, open to complex and changing forces such as military success and failure, ephemeral political alliances and trade contacts of varying intensity. The map carries no date, and is a summary of the traditional settlement patterns and forms detectable in medieval and post-medieval sources whose adapted fossils are still key elements in the fabric of the European scene. Here it is possible to create only a very broad view of the trans-Alpine continent and its peninsulas, and ask what general patterns are visible and what they might mean. Three generalisations can be made: the heart of the land mass is dominated by nucleated settlements, comprising irregular agglomerations of varied size and complexity, but within this general matrix are to be found zones of mixed hamlet and single-farmstead settlement. This core involves western Germany, portions of north-eastern France and lowland England.

Peripheral to this are regions where the settlement forms are regular, geometric, ordered, for example regions throughout eastern Europe, portions of eastern France and northern England, extending also into eastern Jutland and the Danish islands, with a thin specialised band of settlements, often very regular, along the North Sea coasts. Finally, there is an outer rim of regions characterised by single farmsteads and hamlets, to the

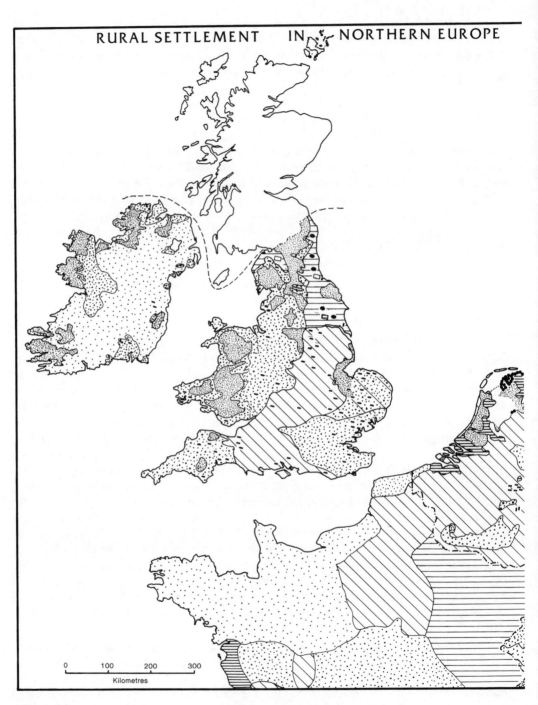

Figure 7.4

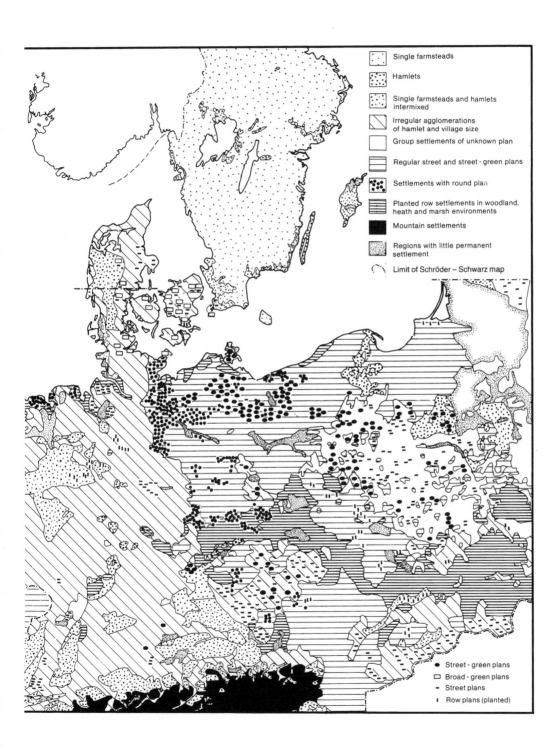

Single farmsteads

Hamlets

Single farmsteads and hamlets intermixed

Irregular agglomerations of hamlet and village size

Group settlements of unknown plan

Regular street and street - green plans

Settlements with round plan

Planted row settlements in woodland, heath and marsh environments

Mountain settlements

Regions with little permanent settlement

Limit of Schröder – Schwarz map

● Street - green plans

▢ Broad - green plans

– Street plans

ı Row plans (planted)

north-west the peninsulas and islands of 'Celtic' Europe, and to the north and east Scandinavia and Slavic Europe. Each of these zones has distinctive settlement characteristics. How can this be explained?

Returning to Figure 3.5 it can be seen that politically Europe expanded from a number of core nodes, regions favoured by productive capacity and location, where state genesis took place. In broad terms, settlement regularity appears to be associated with regions of internal and external colonisation peripheral to these nodes, areas where either the environment at first discouraged settlement, for example woodland and marsh, or where medieval 'European' culture was intruded by military conquest into other cultural zones, notably the Slavic east and the Celtic west. Scandinavia appears different, for there regularity of settlement seems to have been introduced as a result of the conscious decisions by both state and peasantry, rather than imposed by an external force, and settlement regularity is indeed rather more widespread in Sweden than this simplified and experimental map attests. It should be stressed that the forces generating the ring structure involve the migration of ideas as much as people. Flemish colonists were indeed settled in regular villages in eastern Europe and English 'farming folk' were indeed moved from other areas by the Normans into north-west England in the decades after 1092. The regular street and street-green villages they planted are a manifestation of ideas integrated from both the Roman world and the gloom of German forests.

These statements represent vast generalisations, but they may contain grains of truth, and the simple three-zone model created by this description, of core, periphery and outer rim, can, perhaps, be carried to other continents.

Rural settlement in India

The maps of Indian rural settlement which form the basis of Figure 7.5 are derived from the extraordinarily detailed body of data collected by the Indian government, an action which shows how far the rural settlement base is an integral part of the social and economic life of this sub-continent. Two overviews have been selected: the first (Figure 7.5, upper) is a generalised view of settlement types, so far as is possible using categories which are compatible with those for Europe, while the second (Figure 7.5, lower) is a simplified version of data bearing on the holistic structure of Indian rural settlement, and shows the number of villages and their size measured by population (Schwartzberg 1978: 131). In fact, the great majority of Indian country folk live in villages, and, as the map shows, the Himalayan zone is the only area dominated by dispersal; elsewhere, even in the hills, the normal unit of settlement is the small hamlet rather than the homestead. Even in the great delta of the Ganges–Brahmaputra, the hamlets and isolated steadings are present in such numbers that they form settlement chains or 'swarms', to apply a term developed by German scholars to describe this process in parts of Africa. What can be said about this continental scale distribution?

Beginning with description, it is clear that the hill and mountain lands to the north-west, north and north-east, where they are not dominated by dispersed hamlets and single farmsteads, have an admixture of villages, and while cultural variations cannot be excluded, the availability of actual or potential arable land coupled with access to grazings must be a general limiting or enhancing factor. Much of the remainder of the sub-continent is dominated by villages, but the evidence shows that these vary in character. A first great region, embracing much of the Indus Valley and its peripheries,

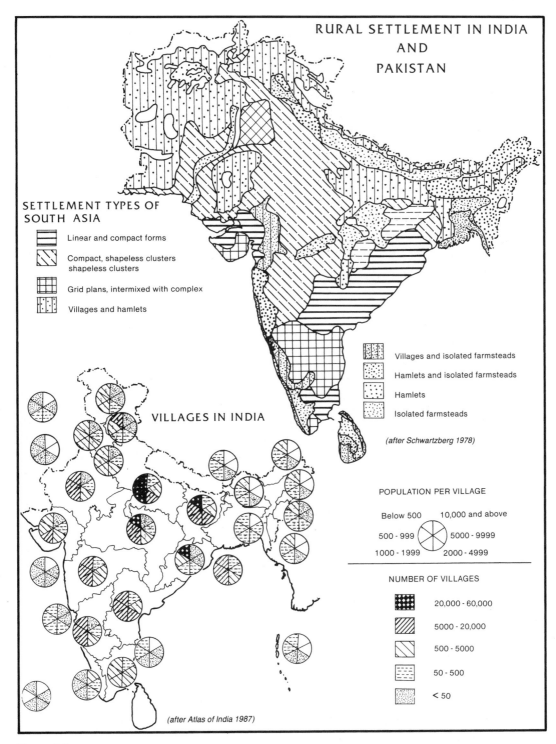

RURAL SETTLEMENT IN INDIA
AND
PAKISTAN

SETTLEMENT TYPES OF
SOUTH ASIA

Linear and compact forms

Compact, shapeless clusters
shapeless clusters

Grid plans, intermixed with complex

Villages and hamlets

Villages and isolated farmsteads

Hamlets and isolated farmsteads

Hamlets

Isolated farmsteads

(after Schwartzberg 1978)

VILLAGES IN INDIA

POPULATION PER VILLAGE

Below 500   10,000 and above

500 - 999   5000 - 9999

1000 - 1999   2000 - 4999

NUMBER OF VILLAGES

20,000 - 60,000

5000 - 20,000

500 - 5000

50 - 500

< 50

(after Atlas of India 1987)

Figure 7.5

and extending southwards to the plateau of the Deccan, is dominated by compact shapeless clusters, with hamlets being rare. To the east, down the Ganges Valley, hamlet intermixtures and eventually mixes with isolated farmsteads appear. To the west in Gujerat, and to the south-east in Orissa and Andhra Pradesh, linear assemblages occur, often growing into more complex forms and possessing outlying hamlets, while in the deep south (Karnatka, extending into Tamil Nadu, where linear forms reappear) is a zone of compact, often square or linear nuclei, with outlying hamlets, these also appearing in southern Gujerat. It is tempting to impose on this map the same general structure as that seen in Europe, a core of irregular agglomerations, surrounded by a peripheral ring of more linear forms, with irregularity and dispersion at the outer limits. There are dangers in doing this, because in Europe the second, regular colonising ring, can be documented and dated; it is probably secondary, i.e. temporally later, than the apparently irregular agglomerations at the core. However, the perceived 'irregularity' in the 'cores' of both continents may simply result from sustained population increases. These affect earlier levels of long-lasting settlements, which may have contained elements of regularity and through adaptation, destruction and rebuilding generate irregularity. To 'explain' the distribution of settlement characteristics in either Europe or India in a few sentences would be unwise. What is wholly clear is that while physical factors cannot be ignored, any fundamental levels of explanation must be cultural, i.e. the presence or absence of external influences, the movements of people and ideas, the local and regional development of political entities, states, and the temporal duration of key factors.

In 1961, taking just the rural population (then approximately 360 million out of a total of 440 million), nearly 50 per cent lived in villages of between 500 and 2,000, 31 per cent in villages with more than 2,000, and only 21 per cent in even smaller settlements, but of course on this scale much hangs on details of definition. In the lower portion of Figure 7.5 the circle diagram for each state is divided into six segments, each segment being assigned to a population classification, from 10,000 and above, to below 500. The shading of each segment shows the numbers of villages in the population classification of the segment, here identifying five categories. The very large numbers of small and smaller villages, 40,000–60,000, may be noted, concentrated in the middle Ganges and regions to its south. Rather large villages, of 5,000–9,999 people, and very large villages, in excess of 10,000, are rarely dominant in the structure of any state's settlement system. It is clear that the general size range is between 500 and 5,000 folk and between 500 and 20,000 village units per state. These figures are cited to give a small measure of scale: in England a figure in excess of 10,000 units is the current estimate for the actual number of villages and hamlets, so that the numbers involved in the Indian sub-continent are almost beyond grasp. It would be fascinating to have comparable figures for China. Collated figures provide a stronger picture: in 1961 there were about 575,000 villages in India, of which 1,835 had over 10,000 people each, 7,228 had over 5,000, and about 282 had less than 500. Kerala accommodates the largest number of villages with populations over 10,000, having 905 or 49.4 per cent of the total, possibly due to population pressures in that small narrow strip of land along the south-western tip of the great peninsula, where villages are merging to become agro-towns.

Rural settlement in Africa

The map of rural settlement in tropical Africa in Figure 7.6 was created using the broad assumption that all nucleated clusters containing over ten compounds were villages while smaller nucleations are hamlets (Davies 1973: 28). Those areas without a significant tendency towards clustering into true nucleations were classified as 'dispersed', no matter how intense the actual pattern of dispersion. This includes those regions of such sustained dispersion, for example in Uganda and eastern Nigeria, that they are described

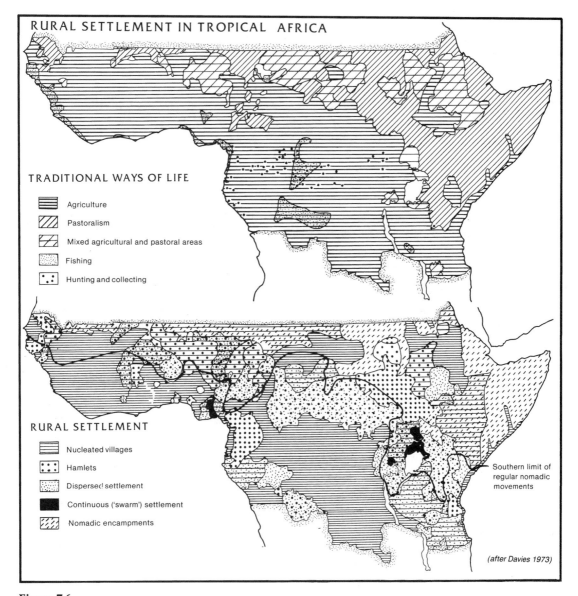

Figure 7.6

as being of 'continuous settlement' (in German, *Schwarmsiedlung*, 'swarm-settlement'). Several key points emerge, notably that nucleation is dominant in West and Central Africa, while in East and in eastern West Africa more mixed settlement is the norm, with an emphasis upon hamlets and dispersion. How can this be explained? Figure 7.6 shows that there is a broad, but not absolute, correlation between settlement and the distribution of traditional ways of life (Davies 1973: 27). Rural communities based upon agriculture, growing cereals such as sorghum and millet, or roots such as yams, sweet potatoes and manioc, have a clear tendency to nucleate. Interestingly this normally involves people who are negroid in character. Those regions dominated by pastoralism, with high cattle numbers per head of population, are much more varied in their settlement characteristics and are dominated by Nilotic, Nilo-Hamitic or (in West Africa) Fulani peoples. In some regions, for example Ethiopia, truly mixed economies once prevailed, and this area, significantly, contained much nucleation. In other regions, parts of Sudan, Chad and other desert margins, areas occupied by agricultural peoples are often also occupied for considerable periods by pastoral nomads: thus the map includes a line presenting the southern limit of regular nomadic movements. In northern Nigeria there is a symbiotic relationship between the sedentary farmers and mobile pastoral nomads, giving rise to very complex settlement variations. Thus the Fulani are divided into a more negroid and more sedentary agricultural group and a more nomadic pastoral group, with links with the trans-Saharan Mediterranean world.

In a fascinating graph reproduced as Figure 7.7, Grenzebach (1984) summarises the morphological characteristics of all the forms of rural settlement present within a large area of West Africa comprising some 200,000 square kilometres. The two axes of the grid are respectively the distance between homesteads along the *y* axis, and the population density on the *x* axis: neither scale is exactly quantified. The classification defines a series of settlement patterns on the basis of the forms which make them up. The varied types overlap, but are further divided into categories on the basis of the nucleated/dispersed division, shown as a horizontally shaded near vertical line. There is also a further division between those which are predominantly rural and those which incorporate a mixture of urban and rural elements. The former are landscapes which are socially and economically underdeveloped, dominated by agrarian, predominantly subsistence economies, while the latter are economically more advanced, market-orientated, with more complex social structures and infrastructures, leading to hierarchical differentiation.

Superimposed upon this already complex diagram are three further categories which define the relationships between the settlement types and the field systems, although this threefold grouping expresses only in general outline what is in reality a wide range of variation. This graph is in effect a complex set diagram, categorising settlements and the patterns of which they are a part, and then placing these categories in a logical and overlapping relationship with each other, using the nucleated/dispersed and predominantly rural/rural-with-some-urban, as broad framing structures. Thus far, the diagram is an interesting form of classification embracing a very large sample, but it is more than this, allowing deep insights into processes of change. It reveals the untold complexity present, yet reduces it to structured, manageable proportions. For the whole of the area studied, Grenzebach was able to estimate both the surface area covered by each category of settlement and the proportion of the total population contained within each category. Thus, while single farmsteads cover 10 per cent of the area, they contain only 5 per cent

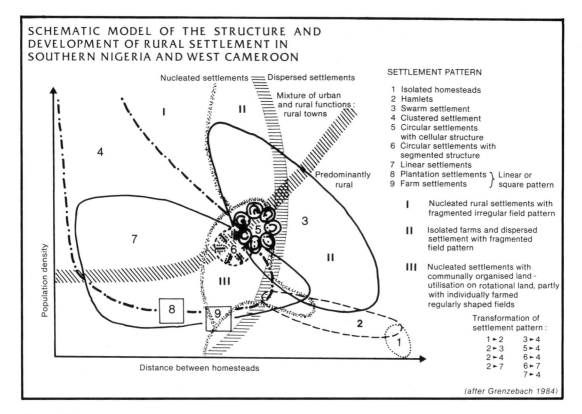

SCHEMATIC MODEL OF THE STRUCTURE AND
DEVELOPMENT OF RURAL SETTLEMENT IN
SOUTHERN NIGERIA AND WEST CAMEROON

Nucleated settlements ⟚ Dispersed settlements

Mixture of urban
and rural functions :
rural towns

Predominantly
rural

Population density

Distance between homesteads

SETTLEMENT PATTERN

1 Isolated homesteads
2 Hamlets
3 Swarm settlement
4 Clustered settlement
5 Circular settlements
  with cellular structure
6 Circular settlements with
  segmented structure
7 Linear settlements
8 Plantation settlements ⎫ Linear or
9 Farm settlements   ⎭ square pattern

I   Nucleated rural settlements with
     fragmented irregular field pattern

II   Isolated farms and dispersed
     settlement with fragmented
     field pattern

III Nucleated settlements with
     communally organised land -
     utilisation on rotational land, partly
     with individually farmed
     regularly shaped fields

Transformation of
settlement pattern :

1 ► 2	3 ► 4
2 ► 3	5 ► 4
2 ► 4	6 ► 4
2 ► 7	6 ► 7
	7 ► 4

(after Grenzebach 1984)

Figure 7.7

of the population, but in contrast *Schwarmsiedlung* (swarm-settlement) involves only 5 per cent of the total area, yet contains 20 per cent of the population, while clustered settlements, *Haufensiedlung,* including the larger towns, account for 5 per cent of the area and 45 per cent of the population. These are interesting figures – even with their limitations – because they are close enough to reality to give a valuable general picture. They allow comparisons and contrasts to be made between areas and point towards more rigorous ways of studying patterns and their development through time.

Further to this, the tabulation by the graph shows how Grenzebach's individual categories are changing, defining what he terms 'the transformation of the settlement patterns'. This idea was used in Chapter 5, Table 5.1, and represents an imaginative leap, recognising that the enclosed defined categories making up the sets in the graph are in fact 'open' to each other, as indeed they are in reality, for regional boundaries are never sharply defined. None of the categories defined is static or unchanging, for each is in a process of 'becoming'. Thus, Type 1, isolated homesteads, are tending to be transformed into Type 2, hamlets; hamlet settlement (Type 2) is tending to intensify to give Type 3, swarm settlements, while overall, there is a clear and sustained tendency towards nucleation. Several key causal factors emerge to account for this: the region saw an eightfold population increase between about 1900 and 1980, from 5 million inhabitants to over 40 million; further, those areas which initially possessed higher levels of population and higher concentrations of settlement were absorbed more readily into the modern process of development and an accelerating process of urbanism than the more sparsely settled areas of dispersed settlement – they are the centres from which innovations and ideas diffuse. The expansion of the network of roads (and railways) has had an additional profound effect on the development of a more matured hierarchy, engendering functional changes in individual places, while the cities are currently growing at an increasing rate, so that the influence of these agglomerated central places will inevitably alter the spatial structures of the purely rural settlement systems. Sparsely settled rural areas, especially those which are difficult to develop with simple equipment because of soil conditions, are the focus of recent processes of immigration, colonisation and settlement – i.e. the areas with plantation and farm settlement, appearing as the intruded rectangles in Figure 7.7 (items 8 and 9). In detail the underlying causes of these changes are complex, but they may be generalised under the one-word heading 'access', for change in rural settlement is normally associated with connectivity, the extent to which outside influences are present or absent.

These three maps, generalising settlement at very large scales, raise interesting research questions; to understand the settlement map of Europe it is essential to invoke historical explanation, but as Grenzebach emphasises in the specific context of West Africa, the dearth of historical sources makes it particularly difficult to determine the spatial–temporal processes of development, i.e. the settlement genesis (Grenzebach 1984: 145–57). Nevertheless, this material, from three continents, can be seen as a step towards showing how the process of classification of on-ground forms and patterns, begun at the scale of the local region or *pays* (Braudel 1989: 127–8), leads directly to the genetic factors underlying the patterns and forms observable at a continental scale, while Grenzebach's generalisations, treating an area somewhat under half the area of Europe, is an indication of what remains to be explored.

References

Adams, I.H. (1976) *Agrarian Landscape Terms: a Glossary for Historical Geography*, London, Institute of British Geographers Special Publication, No. 9.

AJG (Association of Japanese Geographers) (1980) *Geography of Japan*, Special Publication No. 4, Tokyo, Teikoku-Shoin.

Atkinson, F. (1977) *Life and Tradition in Northumberland and Durham*, London, J.M. Dent.

Atzema, O.A.L.C. *et al.* (1990) *Atlas van Nederlanden*: Deel 4, Dorpen s'Gravenhage, SDU uitgeverji.

Barraclough, G. (ed.) (1988) *The Times Atlas of World History*, 3rd edition, London, Times Books.

Barrow, G.W.S. (1981) *Kingship and Unity: Scotland 1000–1306*, London, Edward Arnold.

Bell, C. and Bell, R. (1972) *City Fathers*, Harmondsworth, Penguin.

Beresford, M. and Hurst, J.G. (eds) (1971) *Deserted Medieval Villages*, London, Lutterworth Press, Fig. 17.

Best, R.H. and Rogers, A.W. (1973) *The Urban Countryside*, London, Faber and Faber.

Billington, R.A. (1960, reprinted 1963) *Westward Expansion: a History of the American Frontier*, New York, Macmillan.

Blake, G.H. (1969) 'The Origins and Evolution of Israel's Moshav', *Kulturgeografi*, 109, 293–311.

Blume, H. (1975) *USA: Eine Geographische Landeskunde I*: Der Grossraum in Strukturellem Wandel, 9/1, Darmstadt, Wissenschaftliche Buchgesellschaft.

Bobeck, H. (1937, reprinted 1962) 'The Main Stages in Socio-Economic Evolution from a Geographical Point of View', in P.L. Wagner and M.W. Mikesell, *Readings in Cultural Geography*, Chicago and London, University of Chicago Press.

Braudel, F. (1966, 1972) *The Mediterranean and the Mediterranean World in the Age of Philip II*, London, Fontana, Collins Publishing Group (2nd revised edn 1966; first English edn 1972).

Braudel, F. (1989) *The Identity of France I: History and Environment*, London, Fontana Press.

Bromwich, R. (trans.) (1985) *Dafydd ap Gwilym: a Selection of Poems*, Harmondsworth, Penguin.

Brookfield, H. (1973) *The Pacific in Transition: Geographical Perspectives on Adaptation and Change*, London, Edward Arnold.

Brown, R.H. (1948) *Historical Geography of the United States*, New York, Harcourt, Brace and World, Inc.

Brunskill, R.W. (1985) *Traditional Buildings of Britain*, London, Victor Gollancz in association with Peter Crawley.

Buchanan, R.H., Jones, E. and McCourt, D. (1971) *Man and His Habitat*, London, Routledge and Kegan Paul.

Bulmer, M. (ed.) (1978) *Mining and Social Change*, London, Croome Helm.

Burk, M. and O'Hare, G. (1984) *The Third World*, Edinburgh, Oliver and Boyd.

Burl, A. (1987) *The Stonehenge People*, London, Barrie and Jenkins.

Butzer, K.W. (1964) *Environment and Archaeology*, London, Methuen.

Bylund, E. (1960) 'Theoretical Considerations Regarding the Distribution of Settlement in Inner North Sweden', *Geografiska Annaler*, 42, 225–49.

Calvino, I. (1979) *Invisible Cities*, London, Picador, Pan Books Ltd.

Carey, J. (ed.) (1987) *The Faber Book of Reportage*, London, Faber and Faber.

Carter, H. (1972) *The Study of Urban Geography*, London, Edward Arnold.

Carter, H. (1989) *National Atlas of Wales*, Aberystwyth, University of Wales Press (Chapter 8.5 by C. Thomas, 'Rural Settlement').

Champion, T., Gamble, C., Shennan, S. and Whittle, A. (1984) *Prehistoric Europe*, London, Harcourt Brace Jovanovich.

Chapelot, J. and Fossier, R. (1985) *The Village and House in the Middle Ages*, London, Batsford.

Chorley, R.J. and Haggett, P. (1967) *Models in Geography*, London, Methuen.

Claessens, B. and Rousseau, J. (1987) *Bruegel*, London, Bracken Books.

Clarke, J.I. (ed.) (1966, 2nd edn 1969) *Sierra Leone in Maps*, London, University of London Press.

Cloke, P. (1979) *Key Settlements in Rural Areas*, London, Methuen.

Cole, J.P. and King, C.A.M. (1968) *Quantitative Geography*, London, John Wiley.

Collins, D. (ed.) (1975) *The Origins of Europe*, London, George Allen and Unwin.

Conklin, H.C. (1980) *Ethnographic Atlas of Ifugao: Environment, Culture and Society in Northern Luzon*, New Haven and London, Yale University Press and American Geographical Society.

Conzen, M.P. (ed.) (1990) *The Making of the American Landscape*, Boston, Unwin Hyman.

Craig, J. (1987) 'An Urban–Rural Categorization for Wards and Local Authorities', *Population Studies*, 47, 6–11.

Cunliffe, B. (1988) *Greeks, Romans and Barbarians*, London, Guild Publishing.

Dale, W.L. and Kinloch, R.F. (eds) (1963/4) 'Studies in the Geography of South-east Asia', *The Journal of Tropical Geography*, 17 and 18.

Darley, G. (1978) *Villages of Vision*, London, Paladin.

Davies, H.J.R. (1973) *Tropical Africa: an Atlas for Rural Development*, Cardiff, University of Wales Press.

Dewdney, J.C. (ed.) (1970) *Durham City and County with Teesside*, Durham, British Association for the Advancement of Science.

Dirsztay, P. (1978) *Church Furnishings*, London, Routledge and Kegan Paul.

Dodgshon, R.A. (1987) *The European Past: Social Evolution and Spatial Order*, London, Macmillan.

Dodgshon, R.A. and Butlin, R.A. (eds) (1990) *An Historical Geography of England and Wales*, 2nd edition, London, Harcourt Brace Jovanovich.

Doxiadis, C. (1968) *Ekistics: an Introduction to the Science of Human Settlements*, London, Hutchinson.

Durham County Council (1951) *County Development Plan 1951: Written Statement*, Durham, Durham County Council.

Dussart, F. (1971) *L'Habitat et les paysages ruraux d'Europe*, Liège, Les Congrès et Colloques de l'Université de Liège, 58.

Evans, J.G. (1975) *The Environment of Early Man in the British Isles*, London, Paul Elek.

Everitt, A. (1986) *Continuity and Colonization: the Evolution of Kentish Settlement*, Leicester, Leicester University Press.

Everson, J.A. and FitzGerald, B.P. (1969) *Concepts in Geography: Settlement Patterns*, London, Longman.

Fenton, A. and Walker, B. (1981) *The Rural Architecture of Scotland*, Edinburgh, John Donald.

Forde, C. Daryll (1934, repr. 1956) *Habitat, Economy and Society*, London, Methuen.

Frazer, D. (1969) *Village Planning in the Primitive World*, London, Studio Vista.

Gelling, M. (1978) *Signposts to the Past*, London, J.M. Dent.

Gelling, M. (1984) *Place-Names in the Landscape*, London, J.M. Dent.

Gellner, E. (1988) *Plough, Sword and Book*, London, Paladin.

Glacken, C.J. (1967) *Traces on the Rhodian Shore*, Berkeley, University of California Press.

Gnielinski, S. von (ed.) (1972) *Liberia in Maps*, London, University of London Press.

Goody, J., Thirsk, J. and Thompson, E.P. (1978) *Family and Inheritance*, Cambridge, Cambridge University Press.

Gottman, J. (1961) *Megalopolis: the Urbanized Seaboard of the United States*, Massachusetts, Massachusetts Institute of Technology Press.

Greene, K. (1986) *The Archaeology of the Roman Economy*, London, B.T. Batsford.

Grees, H. (1975) *Ländliche Unterschichten und ländliche Siedlung in Ostschwaben*, Tübingen, Tübinger Geographische Studien, Heft 58.

Grees, H. (ed.) (1986) *Dorfentwicklung Kiebingen: Ortliches Entwicklungskonzept*, Stadt Rottenburg am Neckar, Germany.

Grenzebach, K. (1984) *Siedlungsgeographie-Westafrika, Africa Kartenwerk*, Beiheft W9, Berlin and Stuttgart, Gebrüder Borntraeger.

Guidoni, E. (1975) *Primitive Architecture: History of World Architecture*, Milan, Faber and Faber/Electra.

Hagerstrand, T. (1952) 'The Propagation of Innovation Waves', *Lund Studies in Geography*, Series B, IV, Lund, University of Lund.

Haggett, P. (1965) *Locational Analysis in Human Geography*, London, Edward Arnold, Fig. 4.5.

Haggett, P., Frey, A.D. and Frey, A. (1977) *Locational Analysis in Human Geography I: Locational Models*, London, Edward Arnold.

Hamilton, J.R.C. (1956) *Excavations at Jarlshof, Shetland*, London, Ministry of Public Buildings and Works, Archaeological Reports, No. 1, Her Majesty's Stationery Office.

Hansen, V. (1959) 'The Danish Village: Its Age and Form', *Geografisk Tidsskrift*, 58.

Harley, J.B. (1964a) 'The Settlement Geography of Medieval Warwickshire', *Transactions and Papers of the Institute of British Geographers*, 34, June 1964, 115–30.

Harley, J.B. (1964b) *The Historian's Guide to Ordnance Survey Maps*, London, National Council of Social Service.

Harley, J.B. (1975) *Ordnance Survey Maps: a Descriptive Manual*, London, Her Majesty's Stationery Office.

Hastrup, F. (1964) *Danske Landsbytyper*, Skrifter Fra Geografisk Institut ved Arhus Universitet, 14.

Hedges, J.W. (1984) *Tomb of the Eagles*, London, John Murray.

Hemming, J. (ed.) (1985) *Change in the Amazon Basin II: The Frontier After a Decade of Colonisation*, Manchester, Manchester University Press.

Hodges, R. (1982) *Dark Age Economics: the Origins of Towns and Trade, AD 600–1000*, London, Duckworth.

Homans, G.C. (1941) *English Villagers of the Thirteenth Century*, New York, Russell and Russell.

IDG (Documentation Centre for the Geography of the Netherlands) (1977) *Pictorial Atlas of the Netherlands*, The Hague and Utrecht, Geographical Institute of the State University.

IDG (1979) *Compact Geography of the Netherlands*, The Hague and Utrecht, Geographical Institute of the State University.

Isaac, E. (1970) *Geography of Domestication*, New Jersey, Prentice-Hall.

Johnson, A. (1983) *Roman Forts of the 1st and 2nd centuries AD in Britain and the German Provinces*, London, Adam and Charles Black.

Jones, G. (1968) *The History of the Vikings*, London, Oxford University Press.

Layton, R. (ed.) (1989) *Who Needs the Past?*, London, Unwin Hyman.

Leighley, J. (1963) *Land and Life*, Berkeley and Los Angeles, University of California Press.

Lienau, C. and Uhlig, H. (1967) *Flur und Flurformen* (with English translation *Types of Field Patterns*), Giessen, Kommissionsverlag W. Schmitz.

Lowbury, E. and Young, A. (eds) (1985) *The Poetical Works of Andrew Young*, London, Secker and Warburg.

Lowe, J. (1985) *Welsh Industrial Workers' Housing, 1775–1875*, Cardiff, National Museum of Wales.

MacMaster, D. (1975) 'Geography of Rural Settlements', in J. Clarke (ed.) *An Advanced Geography of Africa*, Amersham, Hulton Educational Publications.

Maitland, F.W. (1897, 1960) *Domesday Book and Beyond*, London, Collins, The Fontana Library.

Mayhew, A. (1973) *Rural Settlement and Farming in Germany*, London, B.T. Batsford.

Mead, W.R. and Jaatinen, S.H. (1975) *The Aland Islands*, Newton Abbot, David and Charles.

Mielke, H.W. (1989) *Patterns of Life*, Boston, Unwin Hyman.

Mills, D.R. (1980) *Landlord and Peasant in Nineteenth Century Britain*, London, Croom Helm.

Mitchell, C. (1973) *Terrain Evaluation*, London, Longman.

Momensen, J.H. and Townsend, J. (eds) (1987) *Geography of Gender*, London, State University of New York Press and Hutchinson.

Morgan, W.T.W. (1983) *Nigeria*, London, Longman.

Mumford, L. (1961) *The City in History*, Harmondsworth, Penguin.

Muter, W. Grant (1979) *The Buildings of an Industrial Community: Coalbrookdale and Ironbridge*, London and Chichester, Phillimore.

Muthesius, S. (1982) *The English Terraced House*, New Haven and London, Yale University Press.

Nationalmuseet (1978) *Historiske Huse i Nordby på Samsø*, Udgivet af Nationalmuseet i samarbejde med Samsø Museum.

Nitz, H.-J. (1987) *The Medieval and Early-Modern Rural Landscape of Europe Under the Impact of the Commercial Economy*, Göttingen, Department of Geography, University of Göttingen.

Oliver, P. (1987) *Dwellings: the House Across the World*, London, Phaidon.

Oliver, P. (ed.) (1971) *Shelter in Africa*, London, Barrie and Jenkins.

OPCS (Office of Popluation Censuses and Surveys) (1991) 'First Results from the 1991 Census', *Population Trends*, 65, 1991, 21–3.

OU (Open University) (1970) *Understanding Society: Readings in Social Sciences*, The Open Press, Macmillan.

Owen, A. (1841) *Ancient Laws and Institutions of Wales*, London, Record Commissioners.

Pacione, M. (1984) *Rural Geography*, London, Harper and Row.

Pennar, M. (1988) *Taliesin Poems: New Translations*, Lampeter, Llanerch Enterprises.

Penoyre, J. and Penoyre, J. (1978) *Houses in the Landscape: a Regional Study of Veracular Building Styles in England and Wales*, London, Faber and Faber.

Piggott, S. (1965) *Ancient Europe*, Edinburgh, Edinburgh University Press.

Piggott, S. (1983) *The Earliest Wheeled Tranport*, London, Thames and Hudson.

Polanyi, K. (1977) *The Livelihood of Man*, London, Academic Press.

Pounds, N.J.G. and Ball, S.S. (1964) 'Core-Areas and the Development of the European States System', *Annals of the Association of American Geographers*, 54.1, 24–40.

Ramm, H., McDowell, R.W. and Mercer, E. (1970) *Shielings and Bastles*, London, Her Majesty's Stationery Office.

Rapoport, A. (1969) *House Form and Culture*, New Jersey, Prentice-Hall.

Rees, A. and Rees, B. (1961, repr. 1989) *Celtic Heritage*, London, Thames and Hudson.

Renfrew, C. (1987) *Archaeology and Language*, London, Jonathan Cape.

Riedé, L. (1983, 1987) *Holland from the Air*, Amsterdam and Brussels, Reader's Digest.

Roberts, B.K. (1987a) *The Making of the English Village*, London, Longman.

Roberts, B.K. (1987b) *Rural Settlement*, London, Macmillan Education Ltd.

Roberts, B.K. (1990) 'Rural Settlement and Regional Contrasts: Questions of Continuity and Colonization', *Rural History*, 1, 51–72.

Roberts, B.K. and Glasscock, R.E. (1983) *Villages, Fields and Frontiers: Studies in European Rural Settlement in the Medieval and Early Modern Periods*, Oxford, British Archaeological Reports, International Series 185.

Roberts, K. (1971) *Bruegel*, Oxford, Phaidon.

Robertson, R.G. (ed.) (1966) *People of Light and Dark*, Ottawa, Department of Indian Affairs and Northern Development.

Robinson, (1984) *People on Earth*, London, Longman.

Rogers, P. (ed.) (1971) *A Tour Through the Whole Island of Great Britain*, by Daniel Defoe, Harmondsworth, Penguin.

Rostow, W.W. (1960, 1971) *The Stages of Economic Growth*, Cambridge, Cambridge University Press.

Ruthenberg, H. (1980) *Farming Systems in the Tropics*, Oxford, Clarendon Press.

Salter, C.L. (1971) *The Cultural Landscape*, Belmont, California, Duxbury Press.

Sauer, C. (1952) *Agricultural Origins and Dispersals*, New York, The American Geographical Society.

Schröder, K.H. and Schwarz, (1978) *Die ländlichen Siedlungsformen in Mitteleuropa*, Forschungen zur Deutschen Landeskunde, Band 175, Trier, Universität Trier.

Schwartzberg, J. (ed.) (1978) *A Historical Atlas of South Asia*, Chicago and London, University of Chicago Press.

Singh, R.L. (1972) *Rural Settlements in Monsoon Asia*, Varanasi, National Geographical Society of India.

Skansen (1974) *Skansen: a Short Guide for Visitors*, Stockholm, A.-B Trycksaker.

Skinner, S. (1982) *The Living Earth Manual of Feng-Shui: Chinese Geomancy*, London, Arkana, Penguin Group.

Smaal, A.P. *et al.* (1979, 1982) *Looking at Historic Buildings in Holland*, Netherlands, Bosch and Keuning nv.

Smyth, W.J. and Whelan, K. (eds) (1988) *Common Ground: Essays on the Historical Geography of Ireland*, Cork, Cork University Press.

Snape, R.H. (1945) *England: a Social and Economic History, Book II, 1066-1300*, London, George Philip.

Spate, O.H.K. (1957) *India and Pakistan: a General and Regional Geography*, London, Methuen.

Spratt, D.A. and Harrison, B.J.D. (eds) (1989) *The North York Moors*, Newton Abbot, David and Charles.

Stamp, L.D. (1962) *The Land of Britain: Its Use and Misuse*, London, Longmans Green.

Steeg, A. (1985) *Monumenten Atlas van Nederland*, Zutphen, De Walburg Pers.

Steensberg, A. (1986) *Man the Manipulator*, Copenhagen, The Royal Danish Academy of Sciences and Letters.

Steensberg, A. (1989) *Hard Grains, Irrigation, Numerals and the Rise of Civilisations*, Copenhagen, The Royal Danish Academy of Sciences and Letters.

Survey of Israel (1970) *Atlas of Israel*, Survey of Israel, Ministry of Labour, Amsterdam, Elsevier.

Szulc, H. (1968) 'Studies on the Silesian Village in the Light of Plans from the Beginnings of the Nineteenth Century', *Kwartainik Historii Kultury Materialnej*, XVI.4, 621–39.

Thirsk, J. (ed.) (1967) *The Agrarian History of England and Wales, IV 1500–1640*, Cambridge, Cambridge University Press.

Thorpe, P. (ed.) (1978) *Gerald of Wales; the Journey Through Wales/The Description of Wales*, Harmondsworth, Penguin.

Tuan, Yi-Fu (1974) *Topophilia: A Study of Environmental Perception, Attitudes and Values*, New Jersey, Prentice-Hall.

Ucko, P.J. and Dimbleby, G.W. (1969) *The Domestication and Exploitation of Plants and Animals*, London, Gerald Duckworth.

Ucko, P.J., Tringham, R. and Dimbleby, G.W. (1972) *Man, Settlement and Urbanism*, London, Duckworth.

Udo, Reuben K. (1982, reprinted 1987) *The Human Geography of Tropical Africa*, Ibadan and London, Heinemann.

Uhlig, H. and Lienau, C. (1972) *Die Siedlungen des Ländlichen Raumes*, (with English translation, *Rural Settlements*), Giessen, Lenz-Verlag.

Uldall, K., Michelsen, P., Stocklund, B., Higgs, J. (Translator) and Kirk, F. (1972) *Frilandsmuseet: The Open-Air Museum*, Copenhagen, The National Museum.

UN (United Nations) (1985) *Compendium of Human Settlements Statistics*, New York, United Nations, Department of International Economic and Social Affairs Statistical Office.

Van Bath, B.H. Slicher (1963) *The Agrarian History of Western Europe, 500–1850*, London, Edward Arnold.

Wagner, P.L. and Mikesell, M.W. (1962) *Readings in Cultural Geography*, Chicago and London, University of Chicago Press.

Wannenburgh, A., Johnston, P. and Bannister, A. (1979) *The Bushmen*, New York, Mayflower Books.

Watson, J.W. and Sissons, J.B. (eds) (1964) *The British Isles*, London, Nelson.

Wolman, M.G. and Fournier, F.G. (eds) (1987) *Land Transformation in Agriculture*, New York, John Wiley and Sons.

Woodforde, J. (1985) *Farm Buildings*, London, Routledge and Kegan Paul.

Woodruffe, B.J. (1976) *Rural Settlement Policies and Plans*, Oxford, Oxford University Press.

Wright, W.D.C. and Stewart, D.H. (eds) (1972) *The Exploding City*, Edinburgh, Edinburgh University Press.

Index

Page references in italic refer to maps and diagrams